Patterns in Evolution

Patterns in Evolution

The New Molecular View

Roger Lewin

**SCIENTIFIC
AMERICAN
LIBRARY**

A division of HPHLP
New York

Cover/text designer: Blake Logan

Library of Congress Cataloging-in-Publication Data
Lewin, Roger
    Patterns in evolution: the new molecular view/Roger Lewin.
        p.   cm.
    Includes bibliographical references (p.   ) and index.
    ISBN 0-7167-5069-4 (hardcover)
    ISBN 0-7167-6036-3 (paperback)
    1. Molecular evolution.   I. Title
QH325.L48   1996
575—dc20

96-25461
CIP

ISSN 1040-3213

Printed in the United States of America.
First paperback printing, 1999

Scientific American Library
A division of HPHLP
New York

Distributed by W. H. Freeman and Company
41 Madison Avenue, New York, NY 10010
Houndmills, Basingstoke RG21 6XS, England

Contents

Acknowledgments

I have been privileged to be present as an observer as the new science of molecular evolution has taken shape, and in my own involvement in it I have benefitted from encouragement and support from many, too numerous to name in full. At the risk of causing offense by omission, however, I should like to express my special thanks to Morris Goodman, David Hillis, and Emile Zuckerkandl.

The Alfred P. Sloan Foundation, showing foresight, set up a program run by Michael Teitelbaum to promote the study of molecular evolution by funding young researchers. I also received financial support from the foundation for the preparation of this book, for which I am grateful.

And to my mother, I give my deepest gratitude and respect.

Patterns in Evolution

Male and female butterflies display the variety and beauty of nature, to which biologists are beginning to apply modern molecular techniques in pursuit of traditional questions of natural history.

A New Window onto Nature

*N*icholas Davies, a zoologist at the University of Edinburgh, Scotland, is a keen observer of animal behavior, one of the best, according to his colleagues. Instead of choosing one of the more glamorous subjects of study, such as chimpanzees or lions, however, Davies has spent the past dozen years watching flocks of dunnocks, or common house sparrows, in the Botanical Gardens of the University of Cambridge. Mundane in appearance they may be, but these little brown birds nevertheless have a social life that is anything but dull. Unlike the legendary monogamy of swans, for example, the social life of dunnocks can take many forms, including one male with one female (monogamy), one male with several females (polygyny), one female with several males (polyandry) and even odd combinations, in which several males may share several females (polygynandry). Fundamental to the evolutionary biologist's credo is that animal behavior is not anarchic, as it superficially seems to be in the dunnocks' case, but is instead shaped by a simple Darwinian rule: namely, individuals seek to maximize their reproductive success, or the number of offspring they produce. In their flamboyant behavior patterns, Davies observed, dunnocks are not indulging in bizarre social extravaganzas, but appear to be playing carefully honed Darwinian games.

Producing offspring is an unequal business in the world of nature, for the interests of males and females are usually different. For males, the Darwinian goal is to sire as many offspring as possible; for females, it is to rear successfully as many of her own offspring as possible. The social structure that emerges for any particular species is strongly influenced by the food resources available and the effort required in harvesting them. In many cases, particularly among mammals, females are able to be the sole providers, in which case they are left holding the baby—literally—with males playing no parental role. The offspring of birds, however, often re-

Dunnocks, or hedge sparrows, have complex mating systems. Genetic analysis has shown that individuals are extremely efficient in shaping their behavior to maximize their Darwinian fitness—that is, to maximize the number of offspring they produce.

quire more food than can be supplied by one provider, so paternal care—typically in the context of monogamy—is common. The variable interplay of these different Darwinian interests is the cause of the complex social lives of dunnocks.

For instance, Davies noticed that even in an ostensibly monogamous pair, the female often tries to solicit mating from a second, or beta, male, and her efforts have the effect of encouraging him to join in the provisioning effort. When this happens, the first, or alpha, male vigorously tries to keep the interloper out. In these cases, it would make Darwinian sense for the beta male to adjust the amount of effort he bestows provisioning the clutch to the level of success he'd had in achieving copulations. Davies noticed that some beta males contributed a lot to provisioning, others less so; clearly, some kind of apparently Darwinian adjustment was going on. For instance, in 80 percent of cases where a beta male managed to mate with a female, it helped provision the clutch. Where beta males failed to mate, they provisioned only 9 percent of the time. Moreover, those beta males that were successful in mating adjusted their provisioning effort in accordance with the degree of their success: the more they had mated, the harder they worked. In other words, the behavior that Davies saw among the birds appeared to equate reasonably well with the Darwinian imperative. But, without being able to identify which offspring had been sired by which males, he could not be certain how well they achieved these goals. His keen eye had yielded great insights into dunnocks' social behavior, but observation alone was insufficient.

In the late 1980s, Davies was presented with a way of determining which offspring belonged to which. Collaborating with Terry Burke, a geneticist at the University of Leicester, England, he collected blood samples from the birds, then, using DNA fingerprinting, matched offspring to fathers. The technique, which had recently been developed at Leicester by Alec Jeffries, identifies unique patterns in an individual's DNA, and shows which individuals are related to which others. The results from the tests showed that the birds were good judges of the appropriateness of their behavior in relation to their reproductive success, but not as good as biologists had once imagined. More frequent mating increased the likelihood of paternity, but the match between mating success and reproductive success (that is, the number of offspring sired) was not exact.

So much for the interlopers' behavior and their Darwinian savvy. What of the females and the alpha males? Davies's earlier observations had revealed that a clutch that is provisioned by two males and the mother is al-

The technique of DNA fingerprinting, developed in England in the mid-1980s, provides a way of identifying individuals and their close relatives. Here, a mixture of DNA fragments is being applied to a gel on which the fragments will be separated, producing a ladderlike pattern unique to the individual.

ways more successful reproductively than a clutch attended to by a monogamous pair. More offspring survive, and they are heavier at fledging than those from monogamously raised clutches. The benefits to the female in encouraging a second male's attention therefore are obvious. Even the alpha male might benefit, too, so long as he sires some 60 to 70 percent of the young, according to Davies's calculations. The typical alpha male's efforts to repel other males suggests that he knows he is unlikely to achieve this critical level of success. Indeed, the DNA fingerprinting data show that, on average, when an alpha male and a beta male provision a single clutch, the alpha male is the father of only between 45 and 55 percent of the offspring. He is therefore right to attempt to keep a second male away.

Davies's insights into the dunnocks' behavior had been hailed as groundbreaking, even before he incorporated a genetic dimension. With the new molecular tools, however, Davies not only sharpened his observations to a degree once considered impossible, but he also became a pioneer in a revolution that is currently rejuvenating and transforming traditional biology. DNA fingerprinting, a technique born from a fortuitous discovery in a molecular biology laboratory, puts a powerful tool into the hand of field biologists asking evolutionary questions that range from identifying an individual's parent (as in the dunnock example) to elucidating the evolutionary relationships of branches of life that have roots billions of years deep. And DNA fingerprinting is but one of a handful of recently developed techniques of molecular biology that either allow access to previously unanswerable questions or make the task of answering them far easier than anyone imagined just a few years ago. The techniques range from comparing the gross structures of proteins to reading the complete DNA sequences of genes and even small genomes. Evolutionary biology, unleashed from earlier technical limitations that allowed biologists to scrutinize only the products of genes—the anatomy of organisms—and not the genes themselves, is launching itself into the twenty-first century, propelled by powerful tools of molecular biology.

An Evolutionary Perspective

"Nothing in biology makes sense except in the light of evolution," the geneticist Theodosius Dobzhansky argued two decades ago. Sweeping though it may be, the statement is fundamentally true, as it must be, for the world of nature is the product of evolution.

For millennia, scholars have striven to identify, describe, and understand the diversity of life, scrutinizing the behavior and anatomical form of organisms in the search for links between them. In pre-Darwinian times, these links were perceived as revealing the pattern of divine creation; with the advent of Darwinian theory, the links were viewed as revealing the pattern of evolution. In the eighteenth and early nineteenth centuries, for instance, scholars saw order in nature in the form of the Great Chain of Being, a concept whose intellectual roots reached back to Aristotle. From the simplest forms of life, the bacteria, to the most complex, *Homo sapiens*, nature was arranged in regularly graded intervals, in a hierarchy that reflected the orderliness of creation. The chain was meant as a description of the world of nature as it always had been since creation, and always would be. When, in 1758, Carolus Linnaeus published his *Systema Naturae*, he was seeking to describe the relationship between organisms with a formal system of classification, a system that is still followed today. By comparing species' anatomy, Linnaeus sought clues to relationships among them, with

In the mid-eighteenth century, Carolus Linnaeus developed a system of classifying species on the basis of anatomical similarity. The system is still used today, with the application of the so-called binomial description of an organism by its genus and species, as in the term *Homo sapiens* that designates modern humans.

Charles Darwin, the father of modern evolutionary theory, had a dream that one day it would be possible to reconstruct a complete tree of life, showing the evolutionary relationships among all living species. This photograph was taken in 1868.

greater similarity implying close relationship, and less similarity more distant relationship. The forelimbs of birds show similarities as wings built on the same fundamental structure, for example, and they are identifiably different from the forelimbs of mammals. These relationships were then reflected in a hierarchical system of classification composed of ever more inclusive groups, going from the individual species themselves as the lowest rung, on through genus, family, order, class, and on up to kingdom. Our own species, for instance, belongs to the genus *Homo (Homo sapiens);* the family is the Hominidae (although, as we will see later, ideas on this are currently shifting in interesting ways); the order is Primates; the class Mammalia; the phylum Chordata; and the kingdom Animalia. There is only one species in the genus *Homo,* for instance, but there are 183 species of primates, and 4000 species of mammal.

With the publication of Charles Darwin's *Origin of Species,* in 1859, and the subsequent acceptance of evolution by natural selection, patterns in nature were recognized as the outcome of common descent, not divine creation. All species display differences that distinguish them, but species with a common ancestor share certain anatomical (and behavioral) features as a mark of their shared history. Broadly speaking, the attempt to recognize these shared features is known as systematics, a pursuit that became the central pillar of traditional biology. Interestingly, the methodology used by biologists in pre- and post-Darwinian times was identical: biologists of both eras compared anatomy in the search for similarities that revealed relationship. What changed was the interpretation of the source of these similarities.

Ultimately, all species are part of a great branching tree of life, the living species being the tips of the twigs, the extinct species being the branches that lead eventually back to a common trunk. In their devotion to systematics, biologists have been clambering around in this tree, seeking to understand evolutionary relationships here and there, collectively piecing together the whole. More than a century ago, Darwin wrote the following to his friend Thomas Henry Huxley: "The time will come, I believe, though I shall not live to see it, when we shall have fairly true genealogical trees of each great kingdom of Nature." Darwin's dream remains just that, a dream. But it is without doubt beginning to take tangible form, with genetic studies playing an important role.

The great natural history museums of the world are a rich legacy of the devoted pursuit of systematics, both before and after Darwin. Biologists fanned out into wild nature, seeking living specimens of the groups of species that interested them, slowly building collections of plants, insects, and higher animals representing the diversity of extant life. They recognized new species and named them in the Linnaean tradition, and discerned relationships to known species. At the same time, paleontologists explored ancient strata, collecting fossils of species that had been part of a diversity of life now extinct, ancestors of modern diversity. It was an endeavor of high adventure, sometimes keen competition and rivalry, and time-consuming laboratory analysis. Sometimes questions were asked at a large scale, such as how the different classes of vertebrates were related to each other; more often, the questions addressed were smaller scale and more tractable, such as the evolutionary relationship among a group of rodents or a branch of arthropods. Discerning evolutionary relationship between organisms based on similarity of anatomy is not as easy as it sounds, because although such similarity may truly reflect common descent, it can also result from organisms becoming adapted to similar environmental circumstances. Fish and whales, for instance, are shaped by the aquadynamic demands of swimming; their physical resemblance is no sign of close evolutionary relationship. Often, discriminating between the two forms of similarity is extremely difficult, and sometimes it is impossible. These difficulties led to many differences of opinion over evolutionary patterns. (One of the great hopes of the use of genetic information was that it would bypass this problem, but it is much more fundamental to the evolutionary process.) Nevertheless, a sketchy outline of life's pattern was assembled, based on traditional methods, but with many question marks. The scope of the vision was enormous.

We share the world with as many as 30 million species, a wondrous diversity of life that is merely the current expression of a continuous process of extinction and speciation. More than 99 percent of species that have ever existed are now extinct, a fact that should make us recognize the rich history from which we derive.

Life originated early in the Earth's history, almost four billion years ago, in the form of simple, single-celled organisms. Across mind-cheating tracts of time, variations on this original theme were played out. Not until a little more than half a billion years ago did more complex organisms, assembled from many cells, evolve. In a burst of diversification aptly called the Cambrian explosion, many different body plans appeared in a brief moment of extravagant evolutionary innovation, each body plan serving as the basic

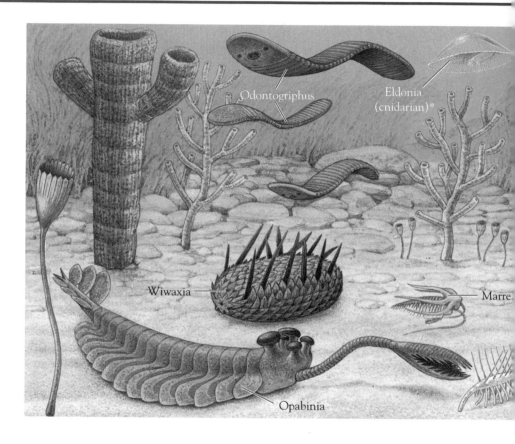

model for multiple species (said to belong to the same overarching group, or "phylum"). Bizarre to our modern eyes, members of these hundred or so founding phyla lived on or near the shallow sea floor. Not all these forms survived, but all (or virtually all) of the living phyla had their origin then. Each has undergone countless rounds of change as evolution continued to unfold around these basic themes, a process that was punctuated by occasional mass extinctions and subsequent bursts of recovery. From a small number of simple organisms in the ocean, nature eventually populated a kaleidoscope of niches in the world, building the communities of species— ecosystems—that we see today.

The ecosystems of the modern world are therefore descendants of four billion years of evolution. They are the ephemeral products of the interac-

The Cambrian explosion 530 million years ago produced many different types of organisms, a few of them depicted here, in a brief geological moment of evolutionary innovation. All of today's phyla came into being, as well many others that did not survive (indicated by asterisks).

tions among members of different kinds of species—among bacteria and fungi, plants and animals, some of which are predators, some prey. Ecosystems are ephemeral, in the sense that change is a constant theme in biology; but permanent, too, in the sense that the same ecological themes are played out again and again through the ages, with different species in similar niches. In ages past saber-toothed tigers were major predators of Africa; now the lion is king. In ages past Deinotheria consumed the woodlands; now the elephant does. Whatever the time in history, whatever the place, individuals of each species are driven by a Darwinian imperative, that of maximizing their reproductive success, as just described with the dunnocks. Each species is engaged in the "struggle for existence," as Darwin put it. The study of ecology—the way individuals of a species and species themselves make their way in the world—is therefore the study of evolution at work, day by day. An important element in this struggle, of course, is the physical world—the chemistry of the soil and the prevailing climate—and the constraints it imposes on nature.

Traditional biology, then, devotes itself to these several issues just mentioned: an understanding and reconstruction of the history of life and an understanding of the behavior of individuals, populations, and species, as components of ecosystems. Achievements in these areas of study have been enormous, especially in the face of the intellectual magnitude of the tasks. During this century, for example, and particularly in the last two decades, field biologists have been studying wild populations of scores of species, an effort that has yielded sophisticated theories of behavior. Despite these achievements, however, traditional biology has typically been regarded as a "soft" science, the intellectual poor cousin to, for instance, particle physics or cosmology, the "hard" sciences, because of the greater mathematical context of the latter (but this is changing, particularly in theoretical ecology). And museum collections, once the frontier of the natural sciences,

For centuries biologists collected samples of the rich diversity of life, such as these shells belonging to a species of variegated tree snail, *Achatinella phaezona*, from Hawaii.

have been increasingly dismissed as anachronistic establishments of little intellectual core, and the curators plying their time-honored trade of systematics as nothing but "stamp collectors." As a consequence, several major museum collections have been dispersed in recent years, to give way to *modern* biology—that is, *molecular* biology.

It is therefore ironic that molecular biology should play midwife to the current renaissance in traditional biology, by offering techniques that allow unprecedented access to genetic information. At the core of every organism is the genetic package, or genome, that gives rise to it. Encrypted within the DNA of the genome are instructions for developing a mature individual from a single, fertilized egg. The means by which an egg becomes a fully formed adult remains one of biology's greatest mysteries and challenges. Also encrypted within the genome of an individual, however, is the history of its species, an unbroken thread of heredity that reaches back to the beginnings of life. It links every species to every other, some closely so, some at great distance, through the number of genes they share. Humans, for example, share genes not only with other mammals, but also with plants, fungi, and bacteria. They will share more genes, though, with their closer relatives than with distant ones. For instance, genetic differences would be small between two geographically separate populations of a species, but much larger between two distinct species with divergent evolutionary paths. The genetic difference between populations of seaside sparrows along the Florida coast is just a fraction of that between, for instance, vervet monkeys and howler monkeys, Old World and New World species that have been evolutionarily separate for perhaps 60 million years.

Biologists have long appreciated that with genetic information to hand, and the ability to analyze such genetic differences, they could address many interesting evolutionary issues, among them the reconstruction of evolutionary trees. Indeed, efforts to obtain such information go back to the turn of the century, when the British biologist George Henry Falkner Nuttall compared the immunological characteristics of certain blood proteins of humans and a few other primates (chimpanzee, gorilla, orangutan, and gibbon). Although he was not explicit about their implications for evolutionary history (because his interest was in blood chemistry), Nuttall's results showed that humans are more closely related to African apes than Asian apes. This was a harbinger of important insights into human prehistory that came six decades later, from experiments by Emile Zuckerkandl and Linus Pauling, of the California Institute of Technology, and Morris

Triplets of nucleotides, or codons, specify individual amino acids in the assembly of proteins. The information encoded in DNA is first transcribed into messenger RNA, a molecule similar to DNA that retains the original DNA's sequence of codons. The messenger RNA is then used as a template for the translation into a protein molecule at the site of protein synthesis, the ribosome.

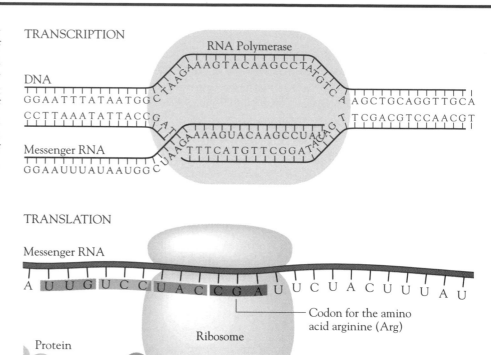

TRANSCRIPTION

RNA Polymerase

DNA

Messenger RNA

TRANSLATION

Messenger RNA

Codon for the amino acid arginine (Arg)

Ribosome

Protein

Arg Amino acid

Goodman, of Wayne State University, also on blood proteins. During those six decades, molecular approaches to traditional biological problems were few, because the appropriate techniques did not yet exist.

Proteins are the product of genes, and so reveal genetic information, although not in the detail given by a complete reading of the gene. The building blocks of DNA are chemical subunits called nucleotide bases, which are of four types—adenine, guanine, cytosin, and thymine (abbreviated A, G, C, and T); strung out like beads on a string, the nucleotide bases form the so-called DNA sequence of genes. Proteins are also assembled like beads on a string, the building blocks being 20 different amino acids. The DNA sequence of a gene directly determines the sequence of amino acids in the protein it produces; a triplet of nucleotides, or codon, in a certain order acts as a code that specifies one of the 20 amino acids. At the very least, therefore, there is three times as much information in the DNA sequence of a gene than in the amino acid sequence of a protein. (In fact, there is more, because certain positions in the triplet can change nucleotide type with no effect on the amino acid for which it codes.) Differences in the properties of proteins therefore reflect differences in the genes of the individuals from which they derive.

Such differences are a consequence of history, created by mutations continuously accumulating in separate lineages. Sometimes a mistake takes place when DNA strands are copied as cells divide: a nucleotide that should have been an A is replaced by a G or a C, or perhaps it disappears altogether. By this or other possible means, mutations arise in the nucleotide sequence of DNA. A mutation that remains uncorrected has the potential to influence a species' evolutionary future, *if* it affects the properties of the protein for which the gene codes. A mutated protein that no longer functions well will be eliminated: individuals carrying the mutation will be at a selective disadvantage compared with others carrying the normal protein. They will leave fewer (or no) offspring, and so the mutation will disappear from the population through natural selection. If the change in the protein's function is minimal, so that the mutation neither impairs individuals who carry it nor imparts a great selective advantage, the mutation may become established at low frequencies in the population. A mutation that enhances the function of a protein, making the individuals who carry it fitter than those carrying the normal form (for instance, by making metabolism more efficient in some way), will soon become common in the population.

Mutations are not restricted to changes in single nucleotides in the DNA sequence. Often, large segments of DNA sequence may be lost or gained; these are known as deletions and insertions. The principle is the same, however. Such mutations may become established as part of a species' genetic package for one of two reasons: either they have no impact on a species' fitness, and are tolerated; or they impart a fitness advantage, and are therefore positively selected for. Many mutations occur, but not all survive to become part of a species' genetic package. The accumulation of such mutations is what comes to differentiate different populations of the same species and, of course, different species. This genetic variation is the stuff of evolution and the raw material that biologists are now able to harvest in order to scrutinize a species' genetic history.

Although the researchers of the 1960s and before had to obtain information on genetic variation indirectly by analyzing proteins, during the late 1970s and 1980s, molecular biologists were gearing up to analyze DNA itself. The first inroad was made when scientists developed the ability to identify changes in certain sequences, known as restriction sites, which could reveal differences in large-scale patterns of DNA in genes from different organisms. Such sites, usually a string of some half-dozen nucleotide bases, are the target of so-called restriction enzymes. Genes exposed to these enzymes are therefore cleaved at these sites, yielding a characteristic pattern of DNA fragments of particular lengths. If a mutation occurs in one or more of these sites, so that the changes block the cutting action, the enzyme treatment will yield fragments of different sizes. Differences in the pattern of fragment lengths in DNA from different individuals of the same species or different species is therefore a clue to mutation, or genetic difference, that has occurred. The technique, called restriction fragment length polymorphism (RFLP) analysis, effectively samples about 5 percent of any gene. The second, and ultimate, assault on genetic history was made possible by the newly acquired ability to read the sequences of genes themselves,

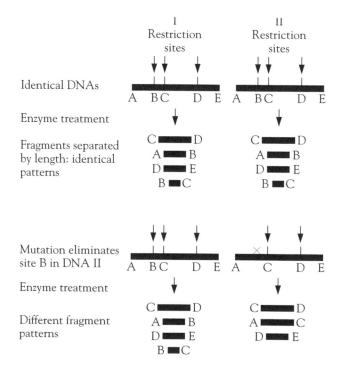

The principle of restriction fragment length polymorphism analysis: Restriction enzymes cleave DNA strands at specific sites, yielding fragments of a characteristic size pattern. Individuals with identical DNA will produce identical fragment patterns (top). If a mutation alters a restriction enzyme cutting site, the fragment pattern will be altered (bottom).

that is, determine the order of nucleotide bases that constitute the gene's string of beads.

The first technique is analogous to being able to identify the existence of paragraphs in a book, but without knowing what they said. The second is like being able to read the entire book, word for word. Initially, DNA sequencing was a slow, costly, and laborious business, but the development of, among other things, automatic sequencing has made the process almost routine. These days a graduate student with no previous experience in molecular biology can learn the sequencing technique and, by the time a thesis is due, can have sequenced a gene of interest in hundreds of individuals, a feat that was possible only within the realms of science fiction just a few years ago.

Impressive though gene sequencing is, by itself it was not sufficient to have a broad impact on the questions of interest in traditional biology. Also needed was a way of plucking desired pieces of DNA from biological material for sequence analysis, and reliable methods of identifying the DNA of individuals. DNA fingerprinting is an example of the latter, and a process known as polymerase chain reaction, or PCR, provided the former. (Both these techniques, developed in the mid and late 1980s, will be described in a later chapter.) So powerful is PCR, and so enormous an impact has it had in molecular and traditional biology, that Kary Mullis, who most fully developed the technique, was awarded a share in the 1993 Nobel Prize for chemistry. With PCR it is possible, in theory, to produce quantities of DNA large enough for sequence analysis, even when the sample material contains just a single molecule of the target DNA. In practice, biologists usually extract DNA from minute amounts of material, such as the root of a single hair plucked from the skin. It is even possible to isolate DNA from dead tissue, which, for the first time, takes genetic analysis into history.

The development of these and other techniques in molecular biology therefore offers traditional biologists a way of obtaining highly informative genetic data. It offers a way of tapping into the history of an individual at all time scales from current (who its parents and siblings are) to recent (how its population became established in its geographical locale) to the most distant (how its species is related to all other species).

New Worlds

The following chapters will provide a glimpse through this newly opened window onto the world of traditional biology. We will see the results of a revolution in action, although one still in its early phases. To set the stage, we will explore how biologists have grappled with the two ways of viewing evolution: through the new perspective of genetic information and the more traditional perspective of anatomy. This "molecules versus morphology" opposition has produced several controversies, notably about human evolutionary history, but harmony is beginning to emerge. We will move on to the use of genetic information in uncovering evolutionary history at several scales of life, including some unexpected insights into the deepest roots of the tree of life itself. The source of genetic variation among populations and species, and the means of discerning it, is the basis of the new revolution; we will discuss this is some detail next, including the notion that genetic information can be used to pinpoint evolutionary events in time, providing a molecular clock.

The final three chapters will show how genetic information is being exploited in ecology and anthropology and applied to bringing museum specimens back to life, not quite literally. Ecologists use the new science to ask why, for instance, green turtles travel thousands of miles to remote breeding grounds and how leafcutter ants became farmers. Anthropologists have gained a deeper understanding of the origin of the human family and of

In laboratories such as this one at California Polytechnic State University, researchers are extracting DNA from dead tissue and even from fossils, raising the unlikely specter of one day reconstructing dinosaurs.

how and when modern humans—people like us—evolved. And museum curators have been able to reach into collections of wolf skins, mummies, and insects in amber to ask questions unthinkable until just recently, each of which has an aura of science fiction about it. The reincarnation of dinosaurs in Michael Crichton's *Jurassic Park* is far-fetched, at least in today's world; but the insights into the lives of individuals and populations long dead, made possible by ancient DNA techniques, still beggar the imagination.

Though still in its early stages, the revolution has nevertheless moved fast. These days it is hard to find a traditional biology department in a university, or a major natural history museum, that does not have facilities for doing DNA analysis. And museum collections, assembled over decades and centuries, are now recognized as priceless repositories of genetic information, not merely warehouses of desiccated skin and bone. Those researchers who had campaigned for the dispersement of such collections as being anachronistic in modern (that is, molecular) biology now recognize their mistake.

Until recently, students who wished to study biology chose either molecular biology or whole organism biology, disciplines divided by a rarely bridged gap. These days, those who study animal behavior or systematics are likely to find themselves learning techniques of molecular biology. And molecular biologists are much more aware that the DNA sequences they read in genes are a record of evolutionary history. Both sides of the old divide are therefore benefitting from the current revolution, but traditional biology is the greater beneficiary. The issues with which it has long been concerned remain the same; and its central methodology—the comparative approach—has not changed. But the new tools of molecular biology add a sharp edge to the way biologists tackle questions they have long wanted answered, but that were beyond their reach.

G enetic analysis solved a stubborn puzzle about the evolutionay relationship of flamingos.

Molecules versus Morphology

A good example of the sort of tantalizing evolutionary question that molecular information appeared to bring within range of solution involves the classification of flamingos, debate about which reaches back to the Greeks. Do flamingos belong with geese (Anseriformes) or with storks (Ciconiiformes)? Flamingos are like storks in having long necks and legs and in some aspects of their basic anatomy, such as in the skull and pelvis. But they are also like geese, with their webbed feet, honking voice, precocious downy young, and the fundamental structure of the bill.

Unable to settle the issue of relationship based on these morphological data, biologists resorted to looking at the feather lice of flamingos. Are the lice like those of geese or storks? (The argument here is that these parasites are highly specialized and usually evolve in close harmony with their hosts.) The answer was that the flamingos' feather lice would make them geese. But the lice were rather *too* similar, given the ancient evolutionary divide that was thought to separate flamingos and geese. In any case, there is the small but real possibility that the flamingos simply picked up goose-type lice as a result of sharing a common ecology, not of sharing evolutionary relationship. Frustrated, some scientists thought the best solution might be a compromise, putting flamingos in a group intermediate between Anseriformes and Ciconiiformes.

In 1985, Charles Sibley and Jon Ahlquist, then at Yale University, declared the conundrum settled. Instead of scrutinizing the birds' anatomy, Sibley and Ahlquist looked to their genetic material using the technique of DNA-DNA hybridization, which effectively compares a large proportion of each species' genes. A species that is genetically close to another species will be more similar in the overall structure of its genes than to a genetically distant species. By this measure, flamingos are storks.

The story of the flamingo's ancestry illustrates two contrasting methods for recovering evolutionary history—the endeavor known as phylogeny. One method—already centuries old—looks for similarities and differences of form and structure, the visible physical characters subsumed under the term morphology. The other method—much more recent—applies the modern techniques of genetic analysis to compare genetic information in species and is known as molecular phylogenetics.

Within the past two decades, molecular phylogenetics has developed from a rare curiosity into a powerful and ubiquitous research tool in modern evolutionary biology. Encouraged by the vast (and ever-growing) flood of molecular data from organisms of all kinds, and by the rise of innovative methods for analyzing these data, as well as the recent development of the necessary computer power, it has been hailed by some as the means to a complete understanding of evolutionary history. Commenting on Sibley and Ahlquist's work on flamingos, Stephen Jay Gould wrote, "We should rejoice in the success of molecular phylogeny," because it provides a method by which evolutionary relationships can be resolved where traditional methods based on comparative morphology fail. "I do not fully understand why we are not proclaiming the message from the housetops," Gould, a Harvard paleontologist, continued. "The problem of phylogeny has been solved."

This example touches on the major issues of phylogeny and the impact on it of molecular techniques. How, for instance, is the evolutionary relationship between organisms to be discerned, whether its analysis is based on morphological or molecular data? What happens when the two forms of analysis yield different conclusions? In other words, is molecular phylogenetics indeed superior to comparative morphology, as Gould implies here?

When molecular phylogenetics first began to make a significant impact, it was undoubtedly viewed by many molecular biologists as a simple, powerful tool able to overcome the problems inherent in traditional methods of classification. (Morphologists were less receptive, as we will see.) Even though there is now no doubt that the new tool on the block is powerful, as expected, it is also clear that its application is nowhere near as simple as was once hoped. This chapter will explore these issues, beginning with a little history. The following chapter will present some examples where molecular phylogenetics has made significant contributions to uncovering the history of life, on many levels.

Homology versus Analogy

The classification of organisms, one of the most ancient practices in biology, has its roots in Greek science. Through comparative morphology—that is, looking for similarities and differences of form and structure—species could be identified as belonging to natural groups, arranged hierarchically. This method had guided Linnaeus in his development of the first comprehensive biological classification (1758), the basis of which remains valid today. Central to recognizing relationship among organisms, however, was the ability to distinguish between two types of physical similarity: homology and analogy. These terms were coined by the British anatomist Sir Richard Owen in the mid-nineteenth century.

Homology, noted Owen, refers to the relationship of "the same organ . . . under every variety of form and function." For example, the human arm, the front leg of a horse, and the wing of a bird are homologous, because they are built from the same basic anatomical structure—that is, the five-digit forelimb of four-legged animals. It is evident, however, that despite their common structural origin, the arm, the horse's foreleg, and the bird's wing serve very different functions. By contrast, analogy refers to the relationship of "a part or organ . . . which has the same function," but derives from different basic body structures. The wing of a bird and of a butterfly are examples of analogous structures: they serve the same function, but the origin of the insect's wing is unrelated to the forelimb structure that generates the bird's wing. The wings of birds and of bats are simultaneously homologous (as structural forelimbs), and analogous (as functional wings).

Sir Robert Owen, the British anatomist who coined the term homology.

21

DNA-DNA Hybridization

Technically, DNA-DNA hybridization is a relatively simple technique for determining the similarity of the full complement of gene sequences between pairs of species. The technique depends on the fact that DNA does not exist as a single strand, but as two, twisted together and bound by hydrogen bonds to form the famous double helix structure. The two strands have complementary sequences of nucleotide bases; A pairs with T, and G with C. (This complementarity is the basis of the molecule's ability to store information and to be readily replicated.) In a single individual, this complementarity is complete, because the two strands are exact matches for each other. In this case, the strength of the binding between the two strands is maximal. When a DNA strand is paired with a second in which complementarity is less than 100 percent, as would be the case with DNA from individuals of different species, the hydrogen bonding between the two strands is weaker. DNA-DNA hybridization measures the strength of bonding that exists between such pairs, and in so doing gives a measure of the mismatch, or dissimilarity, in DNA sequence.

The DNA of individuals to be compared is first extracted from cells and purified, removing the associated RNA and proteins, and then sheared into fragments of about 500 base pairs in

The DNA-DNA hybridization technique effectively compares the entire genetic packages of two species under study. It is based on the fact that strands of DNA with very similar sequences bind together more tightly than those with dissimilar sequences. (See text for details.)

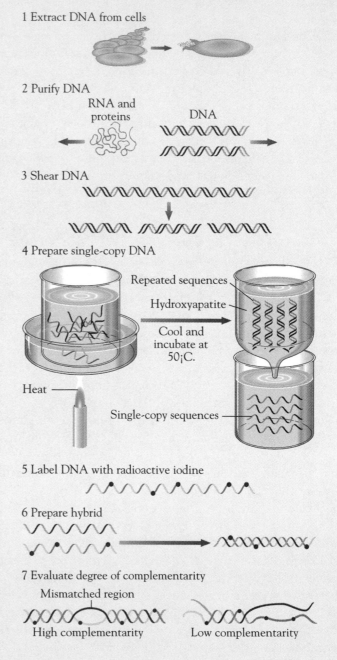

1 Extract DNA from cells

2 Purify DNA

RNA and proteins

DNA

3 Shear DNA

4 Prepare single-copy DNA

Repeated sequences

Hydroxyapatite

Cool and incubate at 50¡C.

Heat

Single-copy sequences

5 Label DNA with radioactive iodine

6 Prepare hybrid

7 Evaluate degree of complementarity

Mismatched region

High complementarity

Low complementarity

length. Much of the DNA of large genomes exists as unique sequences—the genes—interspersed within multiple repeats of short sequences. The most important genetic information is present in the sequences of genes, so that the repeated sequences have to be removed before the two genomes are compared. This is achieved by first heating the fragment mixture to separate the double-stranded fragments, producing a cocktail of single strands; the cocktail is then held at about 50 °C for a short time, during which the repeated sequences rapidly reassociate, leaving the unique sequences separate as single strands; these can then be isolated and compared.

The single-strand mixture of unique sequences from one of the species of a pair to be compared is then made radioactive, with an iodine isotope. The single-strand mixtures from the two species are now mixed together and allowed to hybridize; that is, similar sequences find each other by chance collision, whereupon they will form a double helix. The strength of the binding between two strands is determined by the overall similarity of sequence. When hybridization is complete, the mixture is placed on a hydroxyapatite column, and gradually heated in a water bath that increases the temperature from 60 to 90 °C in increments of 2.5 °C. As the temperature rises, the hydrogen bonds between the hybrids become strained, and eventually break, or melt, yielding single strands again. The bonds melt at lower temperatures in hybrids with many sequence dissimilarities than in those with fewer dissimilarities. Melting is detected when radioactive fragments run off the column as the temperature is increased.

To obtain a measure of dissimilarity, workers may note the temperature at which 50 percent of hybrid molecules have melted, known as the melting temperature, or T_m. Two species, A and B, may be compared by measuring the melting temperatures of, say, a hybrid between identical samples of A (a homoduplex, A:A) and a hybrid between samples of A and B (a heteroduplex, A:B). As a rule of thumb, a difference in melting temperature of one degree is equivalent to a difference in DNA sequence of one percent (that is, one nucleotide in a sequence of one hundred).

The technique yields a measure of the genetic distance between the species, but gives no information about character state (that is, what nucleotide bases are present at any particular position in the genome). For this reason, the method has not enjoyed as much popularity as it might have in molecular phylogenetics, which is methodologically dominated by character-state (cladistic) analysis.

HOMOLOGY
Bat wing

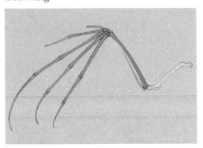

Mouse forelimb

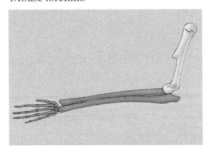

Human arm

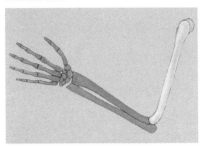

ANALOGY
Bat wing

Bird wing

Butterfly wing

Homology and analogy: A bat's wing, a mouse's forelimb. and a human arm are said to be homologous structures because, although they perform different functions, they derive from the same anatomical elements. Each forelimb has the same fundamental structure but has become modified for different functions. This structural concurrence points to a four-limbed vertebrate ancestor shared by all three species. While the bat's wing and the bird's wing are homologous structures, they are analogous to the butterfly's wing. Analogous structures perform the same or similar functions—in this case flight—but derive from different fundamental anatomical elements shaped by similar environmental demands. The ubiquity of analogy in nature is testimony to the power of natural selection.

Owen was writing in pre-Darwinian times, of course, and his search to discriminate between homologous and analogous structures was a way of discovering what was assumed to be the natural hierarchy of life. Evolution was not yet an issue. Rather, the diversity of life was viewed as the product of God's creation as recorded in Genesis; life on Earth comprised many simultaneous variations on a subset of themes, or archetypes, arranged in a divine order.

With the acceptance of the theory of evolution, biologists' perception of the origin of life's hierarchical order changed—though not their method of discerning it. Evolution made sense of hierarchical order by viewing it as the product of "descent with modification," to use Darwin's phrase. In other words, descendants were anatomically similar to their ancestors, but had been altered in various ways through natural selection: the survival and reproductive success of individuals whose particular modifications fitted them especially well to changing environmental demands. Flattened, paddlelike forelimbs proved advantageous to sea-dwelling mammals (seals and walruses), for example, while other modifications were favored in the forelimbs of mammals who lived by grazing, burrowing, catching flying insects, and so on. In this scheme, archetypes became common ancestors and homology was viewed as an indication of common ancestry, not common design. Analogy was seen as the result not of superficial similarity in a design scheme, but rather of convergent evolution—that is, organisms whose ancestors were not necessarily similar (for example, birds and butterflies) could converge in function (here, flight) by independent paths of evolutionary descent with modification.

Comparative morphology remained, however, the means of recognizing these evolutionary patterns. As the British paleontologist Colin Patterson has noted, "These were changes in doctrine rather than in practice." Comparative morphology had now become a means of reconstructing phylogeny: the genealogy, not of individuals, but of related *groups* of organisms. Darwin was optimistic that, with sufficient time and effort, the history of life could be described in detail. "Our classifications will come to be, as far as they can be so made, genealogies," he wrote in the *Origin of Species*.

During the century that followed this statement, specialists toiled to shed light on genealogies, usually limiting their individual efforts to small segments of the overall tree. Collectively, their efforts assembled a larger picture. Eventually frustration began to permeate this endeavor, however, encouraging a shift of emphasis from phylogeny to genetics, ecology, and the mechanisms of evolution.

The reasons were paradoxical. On the one hand, there developed the belief that the principal shape of the tree of life had already been elucidated, leaving only details to be filled in. On the other, scholars began to recognize that their methods of phylogenetics did not meet the rigors of traditional scientific practice—the criteria used for inferring relationships were not well standardized, for example. Central to this frustration was the need for greater precision in distinguishing between homologous and analogous structures.

Relatively easy to express in words, such a distinction is frequently extremely difficult to achieve in practice, as the flamingo example illustrates. Natural selection can be extremely powerful in shaping anatomy to meet functional needs, so that distantly related species may appear very similar morphologically. The virtually identical appearance of the Tasmanian wolf (a marsupial mammal) and the Eurasian wolf (a placental mammal) is a cogent cautionary tale. The two animals are closely similar to each other in shape and structure, despite the distant evolutionary relationship of marsupials and other mammals. Superficial similarity through analogy is therefore

25

a trickster that can subvert disciplined efforts at inferring phylogeny through comparative morphology. This problem caused some to throw up their hands and declare that the approach of comparing morphologies had achieved as much as was possible. Moreover, many of the deeper (ancient) branches of the tree—the origin and initial radiation of simple multicellular organisms, for example—seemed to remain beyond the reach of traditional comparative methods because, as Patterson has put it, "the clues are poor, the trail is cold."

Phenetics versus Cladistics

In the 1960s two new, independent efforts emerged, both aimed at breaking through this methodological impasse. Although they shared a common goal, philosophically they were extremely different, and there developed an active (sometimes rancorous) debate over which was the more valid. Waves from this debate washed over molecular phylogenetics as it began to offer its first contributions, making for a less than enthusiastic welcome in some circles.

The first of the two new approaches was phenetics. Phenetics, or numerical taxonomy, sought to bring objective judgment to the classification of species, which it terms operational taxonomic units (OTUs). Pheneticists did not explicitly seek genealogies, but instead sought to group species on the basis of *overall morphological similarity*. The hoped-for objectivity lay in the methodology, which demanded that a large number of characters (such as tooth shape in animals and length of petals in plants) be compared by statistical methods among the species under study, with each character given equal weight. The result is a multivariate cluster statistic for pairs of species, ultimately yielding a hierarchy of clusters that represent a description of relationships.

Overall similarity may be a reasonable guide to evolutionary relationships, but phenetics does not seek to reconstruct a phylogenetic tree as such—that is, it does not aim to identify the *branch points* of the tree. If the rate of evolutionary change *within* a tree is uniform or nearly uniform, however, phenetics can offer a measure of the evolutionary distance, or time span, that separates species. Phenetic distance, in effect, measures the *branch length*.

The second approach, cladistics, has its origin in a book published in 1950 by a German entomologist, Willi Hennig, but it did not flower until an English translation appeared in 1966. Unlike phenetic classification, cladistics is explicitly aimed at reconstructing evolutionary histories. For any group of species there is just one true evolutionary history, the history of the group as it actually unfolded. Cladistic analysis seeks to infer that actual history from certain characteristics known as shared-derived traits, or synapomorphies. Those traits that cladists call primitive, by contrast, do not indicate common ancestry. The description of a trait as primitive or as shared-derived depends on context: it may differ according to where in a hierarchy you are looking. An example is the best way to explain this difference.

Willi Hennig, the German entomologist whose 1950 book founded an approach to systematics known as cladistics. These days, cladistics is the most popular technique for reconstructing phylogenies.

Consider the Catarrhini, the infraorder that includes Old World monkeys, apes, and humans. Baboons, chimpanzees, and humans (members of this group) all possess nails on the ends of their fingers, but they are not unique in this respect—all primates have fingernails. Since fingernails were uniquely present in the common ancestor of primates and all its descendants, but not uniquely present in the common ancestor of Catarrhini (because other primates had them), cladistics recognizes fingernails to be primitive characters as far as the Catarrhini is concerned, but designates them as shared-derived characters for primates as a whole. There are a dozen or so characters that uniquely link baboons, chimpanzees, and humans, in relation to New World monkeys; these are shared-derived characters for the Catarrhini. Notice that primitive and shared-derived characters are *both* homologous traits, but they are useful in classification at different *levels* of the evolutionary group (in this case, the primate order).

Without wishing to stray too deeply into the swamps of cladistic classification, it is necessary to ask the obvious question: How is it possible to determine, in a given instance, whether a trait is primitive or derived? The method, known as outgroup comparison, was implicit in the above example: that is, a comparison is sought between the group of species under consideration and one that is relatively close to that group. Consider fingernails again. To decide whether they represent a shared-derived character for the Catarrhini, we examine related groups for their presence. The presence of the character in New World monkeys, for instance, or in prosimians, indicates that it is *not* a share-derived character for the Catarrhini, because New World monkeys and prosimians are both part of the primate order. This methodology leads cladists to claim objectivity for their approach on the grounds that it excludes subjective judgment of relationship based on similarity—and thereby avoids homology-analogy dilemmas. In practice, subjectivity does creep into cladistics, just as it does in phenetics, through the choice of characters for study.

The bottom line for cladistic classification is that actual phylogeny is the ultimate truth, so that recognizing shared-derived characters is the only means of reconstructing a hypothesis of that truth. Phenetic distance measures, being strictly relative, are unacceptable to cladists, whether they derive from molecular or morphological data. Unlike many methods of molecular phylogenetics, the DNA-DNA hybridization that Gould so lauded in the case of the flamingo is a distance measure, not a character-state measure. His enthusiastic remarks ignited a furor among cladists, the reverberations of which inhibited the initial acceptance of molecular phylogeny—and continue still. These days, most biologists recognize the philosophical superiority of the cladistic approach, but this does not mean that distance measures are of no utility in reconstructing phylogeny. One consequence of the phenetics-cladistics war was that it rescued comparative morphology as a means of classification from the doldrums into which it had settled. Thus rejuvenated, comparative morphology was better able to withstand the perceived challenge of molecular phylogenetics.

Ape Affinities

That perceived challenge arose from an early, strong belief on the part of some molecular biologists that molecular data were inherently superior to morphological data—above all, by providing a level of detail (a fundamental record of evolutionary change, written in the base sequence of DNA) that might circumvent issues of homology and analogy, the bugbear of comparative morphology. In addition, molecular and morphological evolution were thought to proceed at very different tempos (the former always regular, the latter always erratic); this was taken to imply that molecular data were more reliable. (There is, however, no simple dichotomy in this respect, as will be explored in a later chapter.) These assumptions had been put to the test in a fascinating and highly charged debate: the phylogenetic history of *Homo sapiens*.

By the mid-1970s, several molecular phylogenetic techniques had been developed. At that time, most of them were used to compare proteins, by electrophoretic, immunological, and sequencing methods, but scientists applying DNA-DNA hybridization could compare genetic material directly. Through this access to fundamental genetic information, "one could hope to obtain a quantitative and objective estimate of the 'genetic distance' between species," wrote Marie-Claire King and Allan Wilson, in a landmark paper in the journal *Science* in 1975. They noted that of all the species that had been examined by these techniques at that time, only two had been subject to all of them: humans and chimpanzees. "A good opportunity is therefore presented for finding out whether molecular and organismal estimates of distance agree," they reasoned.

Genetic information has led to a reclassification of humans and apes. Traditionally (top), humans were the sole members of the family Hominidae, while the great apes (gorilla, chimpanzee, and orangutan) were united in the family Pongidae. Genetically, however, the gorilla and chimpanzee are more closely related to humans than to the orangutan. This relationship is reflected in a new classification scheme (bottom), which puts humans and the African great apes in the family Hominidae, while the orangutan is now the sole member of the family Pongidae .

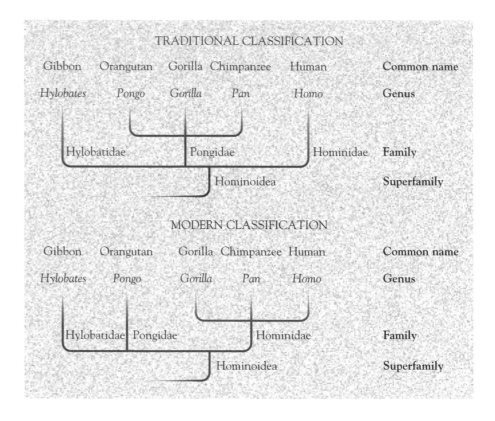

The paper showed that each of the molecular techniques revealed an extremely close genetic relationship between chimpanzees and humans, comparable with that of sibling species. And yet, King and Wilson pointed out, traditional morphological ("organismal") analysis inferred sufficient anatomical difference between the two species to have placed them not only in different genera but in different families. (An unspoken belief that *Homo sapiens* is very different, particularly spiritually and intellectually, from other species had contributed to this classification.) These separate assessments of relationship stood in stark contradiction. Nor was the associated controversy a new one.

Morris Goodman, a biologist at Wayne State University, had effectively established it in the early 1960s. He analyzed the serum proteins of apes and humans (known, collectively, as the hominoids), using an immunological technique to explore genetic affinities among them. The method, simple but effective, relies on the fact that antibodies produced by the immune system are exquisitely sensitive to differences in structure of molecules to which they are exposed.

The first step is to inject human serum albumin (a blood protein) into rabbits. The rabbit's immune system responds by manufacturing antibodies that bind tightly to human albumin. The rabbit's serum can then be collected and used to compare albumins from species related to humans. When human albumin is mixed with the rabbit serum, a strong reaction ensues because of the tight binding, and the albumin precipitates as a solid. Albumins that are increasingly different in protein structure from human albumin will provoke progressively weaker reactions, yielding progressively less precipitant. Using this approach, the pattern of relationships among hominoids that Goodman found was unequivocal: humans and the two African apes (the chimpanzee and gorilla) came together as one group, with the Asian great ape (the orangutan) more distant from them and the Asian lesser ape (the gibbon) more distant still.

We are now well familiar with this pattern, but at the time—and still in most textbooks—a very different perception of relationships was reflected in the formal classification. Humans had been placed in the taxonomic family Hominidae (more colloquially, the hominids), while the African and Asian great apes occupied a separate family, the Pongidae (more colloquially, the pongids). Principally deriving from George Gaylord Simpson's influence, this classification aimed to reflect the adaptive difference between humans and the great apes, but for many it had also come to represent evolutionary history.

Armed with his new data, Goodman proposed in 1962 that the formal classification be changed—that African apes should join humans in the family Hominidae, leaving the orangutan in splendid isolation in its own family, the Pongidae. The idea was not well received. As Simpson put it, "There is not the slightest chance that zoologists and teachers generally, however convinced of man's consanguinity with the apes, will agree on the didactic or practical use of one family embracing both." In other words, the notion that *Homo sapiens* should share the same taxonomic category—that of family—with another organism was simply unacceptable to Simpson and most others. (More than likely, this judgment was not based solely on scientific argument but also included an element of

Homocentrism: the view of humans as special and separate from the rest of nature.)

Goodman's discovery, which David Pilbeam characterized as representing "one of the most significant insights in physical anthropology this century," therefore represented the debut of molecular evidence in the sphere of human origins research. The distress it caused was, however, as nothing compared with the volume of outrage soon to be leveled at Allan Wilson (later King's coauthor) and Vincent Sarich, biochemists at the University of California, Berkeley.

In 1967 they published their investigation of hominoid relationships, using a technique similar to Morris Goodman's, but adding the dimension of time based on the concept of a molecular clock. As will be explained in Chapter 5, such a clock, useful for measuring the time that has elapsed since two species shared a common ancestor, could exist if mutations in the genetic material accumulate at a roughly regular rate: the longer the time

A collection of fossil specimens of *Ramapithecus*. Fossil bones from this apelike creature were first collected in Miocene deposits in Pakistan in the 1930s. Once considered an early human species, *Ramapithecus* is now thought to be ancestral to the modern orangutan, and not related to humans.

of separation, the more mutations will have accumulated independently in the separate species. Relying on this concept, the answer Wilson and Sarich got was that the African apes and humans diverged from each other about 5 million years ago. This diverged sharply from prevailing anthropological thinking, which put the origin of the human family at 15 million years ago, perhaps even as early as 30 million years ago. Such estimates were based on fossil evidence from India of an apelike creature named *Ramapithecus*. Initially discovered in the 1930s, *Ramapithecus* had been designated as the first member of the human family in a major paper published by Elwyn Simons, of Yale University, half a dozen years prior to Wilson and Sarich's report.

There followed a decade and a half of rancorous debate. Wilson and Sarich effectively held that interpretations of anatomy were unreliable as a basis for inferring genealogy, while the anthropologists insisted that comparisons of molecular data were unlikely to produce accurate estimates of the passage of evolutionary time. Both sides had a point. As we have seen, morphologists face great challenges in differentiating between homologies and analogies in anatomical characters. (As it finally turned out, this is precisely what had happened in the erroneous assignment of *Ramapithecus* to hominid status.) Wilson and Sarich, on the other hand, were at that time exploring uncharted territory with their concept of molecular clocks. Traditional biologists were not alone in doubting the reality of clocklike behavior in evolution; some of the sharpest critics were molecular biologists. As a result of these mutual suspicions, the first major collaboration between molecular and traditional approaches in unraveling evolutionary history was anything but smooth.

As the debate unfolded, more and more molecular evidence accumulated—some based on the original immunological technique, but more on newly emerging methods such as protein fingerprinting, protein sequencing, restriction enzyme mapping of DNA, and, eventually, DNA sequencing. There was no significant deviation from the first message: the origin of hominids was a recent event. The figures variously given for that divergence formed a solid range from 4 million to 8 million years ago. No molecular-clock estimate yielded anything like 15 million years, let alone 30 million, no matter what technique was applied.

Not until the beginning of 1982, however, did the anthropological community finally abandon its adherence to *Ramapithecus* as the earliest known hominid. The stated reason with the discovery of a new and espe-

Before the advent of molecular phylogenetic evidence, humans were considered to have diverged from the other apes at least 15 million years ago, based on the identification of *Ramapithecus* fossils as an early human. Since the 1960s, a large body of molecular evidence has accumulated showing that the divergence took place much later, close to 5 million years ago. *Ramapithecus* is now recognized as an ape closely related to the ancestral line of the modern orangutan.

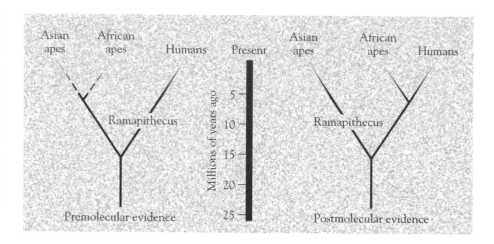

cially complete fossil specimen, *Sivapithecus*, an ape of similar age to *Ramapithecus*. The new fossil showed two things. First, *Sivapithecus* and *Ramapithecus* were very closely related. Second, *Sivapithecus* was related to an ancestor of modern orangutans. If this was true, then *Ramapithecus* must be taken out of the line of ancestry to hominids, because the orangutan group diverged early in the history of the human-ape (hominoid) stock. Molecular phylogenetics had demonstrated its ability, at least in this case, to penetrate the confusion created by the analogy-homology dilemma.

King and Wilson Raise the Stakes

Moving beyond the human interest of this controversy, however, King and Wilson's 1975 survey had powerful resonances for the whole biological enterprise of phylogenetics. Two conclusions flowed from their observations — one explicit, the other implicit. The first was that molecular and morphological evolution could proceed at very different rates, and that the reason for this lay in the nature of molecular evolution. Geneticists had tended to focus on mutation in the structure of proteins as the basis of evolutionary change, and this is certainly correct in many cases. However, argued King and Wilson, more dramatic organismal evolution could come about through mutation in regulatory genes than in structural genes, an argument that Emile Zuckerkandl had proffered some years earlier. Structural genes code for the production of proteins, whereas regulatory genes are responsible for orchestrating the activity of structural genes and, ultimately, for the developmental unfolding from (in higher organisms) a fertilized egg to mature individual. Simple mutations in regulatory genes therefore have the potential for promoting complex changes in development, leading to a very different mature organism. Molecular and morphological evolution may therefore be uncoupled from each other. This conclusion, obvious in hindsight, would have an important impact on the understanding of the evolutionary process.

By this time, there had developed strong support for the notion that mutations at the level of the gene accumulate at a roughly regular rate, providing the basis for a molecular clock. Molecular evolution was therefore considered to proceed in a simple, clocklike manner, while morphological evolution may be anything but clocklike. The obvious implication (the second conclusion from the King-Wilson observations) is that molecular data are going to be much more reliable indicators of evolutionary change and relationships than morphological data.

Molecules, Morphology, and Mice

A decade later, in 1985, the opposition of these two forms of data was made explicit in a major international conference with the title "Molecules versus Morphology." The goal of the conference was to try to assess the power of the two approaches in reconstructing phylogeny.

Participants at the gathering heard arguments that the mode of molecular evolution was beginning to be seen as far more complicated than had been believed, making molecular phylogenetics less reliable than had been assumed. Nevertheless, one important contribution did seem to demonstrate the superiority of molecular methods in a very direct manner.

An inherent difficulty for phylogenetics is that its major goal—that of *knowing* a particular phylogeny—is impossible, because there is no way of testing a particular conclusion. Phylogeny is history, and conclusions about it are at best hypotheses, not certainties. When molecular and morphological phylogeneticists reach their separate conclusions, they are therefore comparing their separates hypotheses, and neither group can be *certain* they are right. If it were somehow possible to do molecular and morphological phylogenetics on a *known* phylogeny, however, then a direct test of conclusions would be possible. Walter Fitch, then of the University of Southern California, and William Atchley, then of the University of Wisconsin, realized that inbred strains of laboratory mice offered such a possibility.

The phylogeny of these mice has been recorded over the past 70 years, as new strains were developed. (A "strain" in this case refers to a genetically distinct subpopulation of a species; it is less distinct than a subspecies, however.) The molecular data were protein variants from 97 gene loci; the morphological data were ten measurements on the lower jaw on the mice at 10 weeks of age. Fitch and Atchley, using five different methods to analyze the two datasets, found that the molecular data accurately reconstructed the known phylogeny, whereas the morphological data did not. They conceded that the mouse phylogeny covered only 70 years—an extremely short period compared with typical phylogenies—and that the morphological data were quantitative (that is, widths and lengths), whereas morphologists typically use qualitative traits (that is, presence or absence of homologous features); both these facts may militate against producing an accurate phylogeny based on morphology. Despite these caveats, Fitch and Atchley concluded that "molecular data appear to be superior to morphological data" for reconstructing phylogenies. Many morphologists argued that these caveats invalidated the comparison. Nevertheless, the experiment had an important impact on the wider assessment of "molecules versus morphology."

Molecules had apparently triumphed. For instance, at the 1985 meeting Sibley and Ahlquist stated that "molecules can reconstruct the phylogeny with a high degree of accuracy." Three years later, Leslie Gottlieb, of the University of California at Davis, drawing on experience with plant phylogenies, concurred: "The molecular data are self-sufficient in that their usefulness does not depend on concordance with other lines of phenotypic evidence."

Homoplasy and Homology among Genes

The initial assumption of the superiority of molecular data for reconstructing phylogenies rested on their being seen as a fundamental record of evolutionary change, rendering the problem of misleading analogy (so-called homoplasy) minimal. Molecular biologists regarded as unlikely the possibility that detailed similarities in the sequence of genes from unrelated species would be the result of chance or selection for such sequences. However, DNA sequences are now known to be much more susceptible to convergent evolution than was supposed, making the homology-analogy conundrum a problem not just for morphological phylogenetics but also for molecular phylogenetics. As Colin Patterson has put it, "Molecular homologies are no more secure, and are possibly more precarious, than morphological ones."

The argument that homoplasy in DNA sequences is a minor issue is very simple. On the one hand, analogous morphological structures evolve in unrelated species because such structures are shaped by natural selection, which is a powerful force for producing functionally and structurally similar characters. By contrast, however, as will be described in detail in the following chapter, natural selection is considered to play only a relatively small role in driving evolution at the molecular level; instead, most changes in DNA sequence are thought to be the result of chance events, a process known as neutral molecular evolution. The conclusion is that, since gene sequences in different species are unlikely to evolve in parallel, they are therefore isolated from the force producing homoplasy.

In fact, analogous gene sequences can arise under neutral evolution, probably at counterintuitively high rates. It would simply require identical mutation at the same site in a gene in different species; given the limited number of states (one of four nucleotides) possible for each site, the chance is quite high that mutations will produce short spans of DNA with similar sequence. Of course, the probability of mutation at any particular site is not great for most genes. Given sufficient time, however, the problem of homoplasy becomes significant. In a review of 42 morphologic and 18 molecular cladistic studies, Michael Donoghue and Michael Sanderson, both of Harvard University, found comparable levels of homoplasy in the two datasets. However, proponents of DNA-DNA hybridization argue that the technique minimizes the problem, because it effectively compares a very large proportion of species' genomes. The number of nucleotide positions at which homoplasy is likely is therefore tiny compared with the number of homologous positions. Statistically, therefore, DNA-DNA hybridization renders homoplasy insignificant.

Setting the problem of homoplasy aside, recognizing homologous gene sequences is by no means simple. Indeed, in some ways it is more complex than for anatomical features. Biologists have spent millennia ruminating over the true meaning, or meanings, of physical homology, and the philosophical discourse continues. But when molecular biologists began using their own data in an evolutionary context, they injected a further level of confusion—in part because of imprecise terminology.

When the sequences of the same gene from two species are compared, they will exhibit a certain degree of similarity: let's say that 50 percent of the sequence is the same. Initially, it was common for molecular biologists to describe these two genes as "50 percent homologous," whereas they simply meant 50 percent *similar*. The two genes might not be homologous at all, in the true sense of sharing a common ancestry. As a result, noted David Hillis, a molecular biologist at the University of Texas, "Molecular biologists may have done more to confound the meaning of the term homology than have any other groups of scientists." The word homology should be restricted to *common ancestry*, and the word similarity to *likeness of sequences*.

Seeking genuine homology between gene sequences is much more difficult than may be imagined. In two species that have recently diverged, there may be no sequence differences at all, or perhaps just a few. Aligning the two sequences side by side will reveal a very high level of sequence similarity, which can be taken as an indication of homology. But with the passage of evolutionary time, the two homologous sequences will independently incur mutations, so that in ancient divergences the level of similarity may become extremely low. It is theoretically possible that, given sufficient time, the level of similarity will be no higher than would have been generated by chance, for the following interesting reasons.

At the molecular level, the functionally important unit of proteins is their three-dimensional structure, which governs the way they interact with molecules such as DNA, RNA, other proteins, carbohydrates, and fats. Protein molecules with very similar three-dimensional structures can be constructed from surprisingly different sequences of amino acids—which are, correspondingly, coded for by different DNA sequences in the operative genes. In the evolutionary context, therefore, the same gene in two diverging species can undergo substantial mutation while maintaining the functional integrity of the proteins it produces. These genes are homologous, despite the dissimilarity of their DNA sequences.

A further complication arises because mutation typically is not restricted to the change of nucleotides at specific sites. Sections of the gene may be lost (deletions), while new sections may be included (insertions). Simply aligning such genes will therefore not immediately reveal their

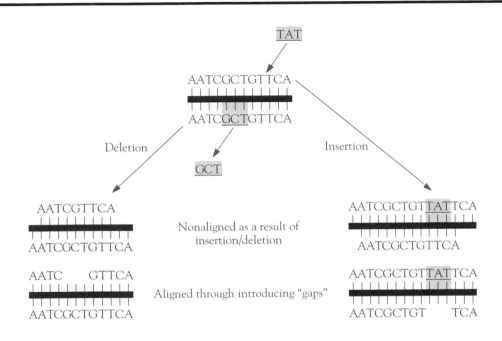

Segments of DNA may become inserted into a gene, or deleted from it, introducing significant dissimilarities in the sequences of related genes. The similarities between the genes are determined by introducing gaps in the known sequences, then sliding the segments (on paper) until they align.

more fundamental sequence similarity. Compensating analytical methods have been developed that allow for "gaps" in the alignment process—effectively sections of the two sequences are slid alongside each other in a search for the islands of similarity. Achieving a correct alignment of sequences in this way is one of the trickiest problems in molecular phylogenetics.

Genes in Different Guises

Until quite recently, the genomes of higher organisms were imagined to follow the simple organization that had been discerned for bacteria. For instance, in bacteria each gene occurs only once—that is, as a single copy. For higher organisms, however, this pattern has turned out not to hold, a fact that makes the notion of homology much more complex in relation to genes than in the case of anatomy.

Many genes in higher organisms exist as multiple copies; collectively, these are known as gene families. Sometimes the members of the gene family are identical to each other, so that the cell can produce large amounts of a single product in a short time. The ribosomal genes that code for the protein-synthesizing machinery are an example. Often, however, the members of the family became different from one another and may perform slightly different roles. For instance, in primates the protein that carries oxygen in red blood cells, globin, exists as a family of half a dozen genes, each a slight variant on the others and each functioning at different periods during the development of an individual, from embryo to adult. Such families arose during evolution by the duplication of an existing gene. Initially, the sequence of the original gene and that of the new one would have been identical. But with the passage of time differences would accumulate, producing the variants mentioned. Some gene families have just two members; others have dozens.

The recognition of gene families prompted molecular biologists to take account of this complexity by coining new terms for different aspects of homology. For genes in which duplication has not occurred (and which therefore exist as single copies), the term orthology was introduced for making comparisons between related species; orthology is functionally equivalent to traditional morphological homology. The term paralogy was introduced to describe the relationship among members of a gene family, such as the globin family, within a species.

The existence of paralogues presents potential pitfalls in molecular phylogenetics that may lead to erroneous conclusions about the evolutionary history of related species. A hypothetical example will illustrate this concept. Imagine an ancestral species that possesses a gene, X. Now suppose that the gene duplicated 10 million years ago, so that the species now has a gene family, X1 and X2. Originally identical in sequence, X1 and X2 will gradually have accumulated independent mutations within the species over the 10-million-year period. Further suppose that the species split 5 million years ago, yielding two daughter species that each contains the two-member gene family. Finally, suppose that a molecular phylogeneticist wishes to infer the evolutionary history of the two daughter species by examining the sequence of gene X, in ignorance of the fact that the gene has two variants. If, through ill luck, gene X1 were to be isolated from one of the daughter species and gene X2 from the second, the inferred history would be distorted. Specifically, a comparison of the two sequences, using the difference as an indicator of the time since evolutionary separation, would imply that the species diverged 10 million years ago—but that is in fact the time of gene duplication, not the time of speciation. Such a conclusion is said to reflect the *gene tree* rather than the *species tree*.

ACTUAL EVOLUTIONARY HISTORY   INFERRED EVOLUTIONARY HISTORY

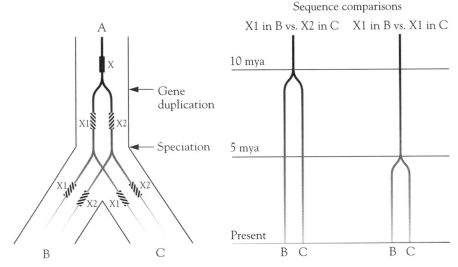

Gene duplication can lead a phylogeneticist to compare the wrong genes and produce an erroneous phylogeny. Here, a gene X in a species A duplicates at 10 million years in the past, yielding a paralogous pair, X1 and X2. The species then splits at 5 million years ago, yielding the daughter species, B and C, each containing the gene pair. If a phylogeneticist compares the sequences of genes X1 in both daughter species, a correct phylogeny is produced; this is the species tree. But if the phylogeneticist compares X1 in species B with X2 in species C, then an incorrect phylogeny is produced. This erroneous phylogeny, which reflects the history of the gene not the species, is the gene tree.

    The terms orthology and paralogy therefore refer to aspects of homology that arise from a gene's history among related species or within a species. A third form of homology, also restricted to the molecular level, is known as xenology. (The Greek root *xeno-* comes from the word meaning "foreign," "strange," or "guest.") Although species' genetic packages are largely restricted to *vertical* inheritance through successive generations, remaining isolated from the genes of other species, occasionally this integrity breaks down. Specifically, genes may be transferred *horizontally* between species, often as passengers on certain viruses, and eventually become permanently incorporated into the host's genome. (A virus must integrate itself into the host cell's genome in order to reproduce; occasionally viral replication "picks up" host genes or "leaves behind" viral genes—or an earlier host's genes.) Common in microorganisms, horizontal transfer is rare

41

Genes in Pieces

Few questions are as fundamental in biology as the origin of the structure of genes of higher organisms. And yet, more than four decades after the elucidation of the basic configuration of DNA, from which genes are built, biologists still debate how these genes came to be the way they are. Moreover, the disagreements are not over mere details, but reveal a deep divide in how scientists interpret the path of evolutionary history at the molecular level. The two sides could hardly be further apart. One views genes of higher organisms as recently developed structures; the other sees them as exhibiting vestigial signals of ancient processes going back to the origin of life itself.

In 1977 the world of molecular biology experienced a major surprise. For the previous two decades or so, biologists had scrutinized gene structure and function in what was assumed to be a model organism, the bacterium *Escherichia coli*. A rule that emerged strongly in this unicellular organism was the existence of a direct correspondence between the structure of genes and the proteins for which they code. The string of nucleotides that made up a gene was translated directly, via a messenger RNA molecule, into a string of amino acids that make up a polypeptide or protein. Biologists expected this one-to-one correspondence established for *E. coli* genes to be ubiquitous in all of higher nature, from orchids to elephants. It turned out, however, that the information that codes for proteins in the genes of

higher organisms is contained in small packets, which came to be called exons, interrupted by stretches of DNA that apparently code for nothing, which came to be called introns. The introns, of which on average there are about half a dozen to a gene, are vastly bigger than the exons—typically, 10 times as long. One of the questions that the discovery of introns prompted is, Why genes in pieces? Another, related, question is, Where did the introns come from?

Since the DNA sequence of introns apparently does not code for proteins, another function had to be sought, perhaps in history and evolution. The Harvard biologist Walter Gilbert quickly suggested that the intron-exon structure would promote evolution: rather than depending on incremental mutations in a gene sequence to produce novel genes, the joining together of different exons in different combinations could quickly produce new, functional genes—especially if a single exon coded for some kind of unit within proteins that was functional in structural or catalytic respects. He called the process exon shuffling. Two other researchers, James Darnell of Rockefeller University and Ford Doolittle of Dalhousie University, Nova Scotia, then suggested that, because there could be little short-term benefit in splitting up previously intact genes by the introduction of introns, the introns must have been present in the earliest, primordial genes and have subsequently been eliminated from the simpler genomes of bacteria.

Not everyone agreed, however. Prominent among the opposition were the British biologists Tom Cavalier-Smith, now at the University of British Columbia, and John Rogers of Cambridge University. They suggested that, since there are many known molecular mechanisms by which introns could be inserted into modern genes, why argue that they were there from the beginning and subsequently pruned away by loss in certain instances? And anyway, the pattern of introns in modern genes is extremely irregular, making the late insertion of introns a more parsimonious explanation. Opinions began to polarize between "introns-early" supporters "introns-late" supporters. (By late is meant within the past billion years.)

One of the most powerful bodies of evidence in favor of introns-late is the pattern of the history of life (phylogeny)—a line of evidence that has been assiduously developed by Jeffrey Palmer and John Logsdon, both of Indiana University. Although introns are present in the genes of higher organisms, they are completely absent in simpler organisms, such as true bacteria and archaebacteria. To explain such a pattern under the introns-early hypothesis, one must invoke the parallel loss of tens of thousands of introns in many independent lineages, including their complete loss in some. This, argue Palmer and Logsdon, is unlikely in the extreme. Much more parsimonious is the proposal that the few lineages that now possess introns do so through their late insertion into previously intact genes. So far, no introns have been found in the genes of bacteria, which may be taken as indicating they never existed in these organisms. However, only a small fraction of genes in known bacteria have been examined, and only the tiniest proportion of all existing bacteria have been studied. The eventual discovery of intron-containing

(continued on page 44)

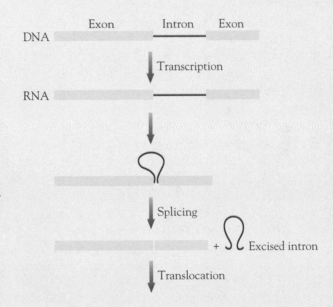

Genes in higher organisms are composed of exons, which contain the code for proteins, and introns, which code for nothing. Once a gene is transcribed into messenger RNA, the introns are spliced out and the exons are united, to form a continuous coding region.

43

genes in some of these organisms would favor the introns-early hypothesis.

Several lines of evidence ought, in principle, to settle the early-versus-late debate. But they don't. For instance, evidence on intron position in genes ought to be decisive. If introns arose at the root of evolution, their positions in the same genes of widely diverged lineages should be the same. If they were inserted late in evolution, after the lineages had diverged from each other, then intron positions would differ, because insertion is implicitly assumed to be a random process. The evidence is equivocal, including some spatial patterning of introns and some apparently random distribution. This may be taken to support the introns-late hypothesis. Nevertheless, the separate processes of intron loss and intron "sliding" could have created the observed pattern.

A crucial feature of the exon theory of genes is the necessary correspondence of exons with some structural feature of proteins. In the early 1980s, Mitiko Gö, of Kyushu University, Japan, identified compact regions of the globin protein that she called modules. She could see where the two known introns divide compact regions in globin, but identified a third such region marked by no intron. Perhaps the third intron had been lost in animal genes, she speculated. Within a year, plant globin (known as leghemoglobin), was sequenced—and shown to contain a third intron, precisely in the position Gö had predicted.

This was the first fulfilled prediction based on the exon theory of genes, proposing modules as the structural units in proteins. A second came a decade later, when an intron was discovered in a gene from mosquitoes coding for the protein triosephosphate isomerase—at the position predicted by Gilbert and his colleagues in 1986. Few such predictions have been fulfilled, however, and in the case of triosephosphate isomerase, more introns have recently been discovered in positions not predicted by the module hypothesis. On the face of it, this may be seen as fatal to the introns-early hypothesis, at least as based on Gö's notion of modules. But although the patterns produced by Gö-type analyses are powerful guides, they are *only* guides, Gilbert says; it is possible to keep breaking the projected structure down, yielding additional putative intron positions. Argument and counterargument continue, with no consensus—in part because the issue centers on deep evolutionary events that, by their nature, are difficult, perhaps impossible, to test experimentally.

among higher animals, but not unknown. A clue that xenology is at play surfaces when a gene in two distantly related species is surprisingly similar in sequence, even though other genes are very unalike.

One of the most important discoveries in molecular biology during the 1970s was that, contrary to expectation, the coding sequence of genes in higher organisms does not exist as a continuous string of nucleotides, as it does in bacteria. Instead, the sequence is split up into significant fragments, called exons, that are separated by long noncoding sequences, called introns. During evolutionary history, genes may have been assembled from a selection of exons from other genes (a process termed exon shuffling). This fact leads to a form of homology impossible in an organism's physical characters. Homology is all-or-nothing in morphology, but exon shuffling allows *partial* homology in molecular evolution. For instance, the gene that codes for tissue plasminogen activator in mammals is made up of a collection of exons from genes that code for other proteins: plasminogen, fibronectin, and epidermal growth factor. This means that the gene for plasminogen activator is partially homologous (strictly, paralogous) with each of these other genes.

During the 1980s, the existence of pseudogenes, or dead genes, was discovered. These are silent copies (minus the introns) of active genes. They pose a potential complication for molecular phylogenetics, as the following scenario, related to our earlier hypothetical illustration, reveal. Suppose gene A in an ancestral species duplicates, to form the paralogues A1 and A2. As before, suppose that the ancestral species forms two daughter species, each of which contains both A1 and A2 genes. Now suppose that in one daughter species gene A1 is silenced—becomes a pseudogene. Simultaneously, in the second species, A2 is likewise silenced. A molecular phylogeneticist, unaware of the actual history of this gene, would unwittingly compare the sequences of these genes as orthologues—not as the

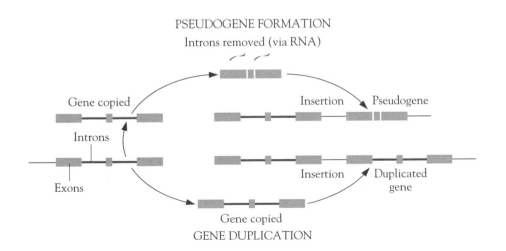

PSEUDOGENE FORMATION
Introns removed (via RNA)

Gene copied

Insertion   Pseudogene

Introns

Insertion   Duplicated gene

Exons

Gene copied
GENE DUPLICATION

Genes may duplicate and form multigene families (bottom). But sometimes a gene may be copied via an RNA intermediate, from which the introns and regulatory sequences are removed (top). Such a pseudogene is therefore nonfunctional.

Pseudogene formation can deceive a phylogeneticist into constructing an erroneous phylogeny. Gene A duplicates, forming A1 and A2, which are passed to two daughter species. If, now, A1 in one species becomes silenced through becoming a pseudogene, and A2 is similarly silenced in the other species, a comparison of gene A in the two species would yield a gene tree, not a species tree.

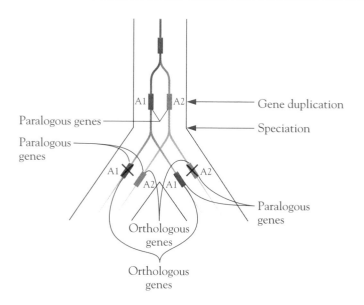

paralogues they really are. He would therefore deduce the gene tree (recording the duplication event that formed the two-member gene family) while imagining mistakenly that he had deduced the species tree (recording the split into daughter species from the ancestral organism).

Without delving further into the minutiae of modern molecular evolutionary research, suffice it to say that these and other discoveries create a picture of the genome of higher organisms as an extremely complicated, extremely dynamic system. This complexity and dynamism have the potential to make molecular phylogenetic data less straightforward and reliable than earlier, more static concepts of the genome suggested—and at least as troublesome as some morphological data.

As an apt postscript to this discussion, it is worth noting that Fitch and Atchley have continued their work on inbred mice, expanding the range of genetic information they scrutinize. By now they have analyzed data from more than 200 genetic loci (from general protein-coding genes, immune system genes, and certain viruses that become incorporated into the mouse genome) in 24 strains of mice. The general protein-coding genes still give a phylogeny consistent with the known evolutionary history, whereas the immune system genes and viral genes do not. "These analyses suggest that not all genetic data have equivalent information content for phylogenetic reconstruction," the authors concluded in 1991.

The message is that information from a single gene, or even many genes, does not necessarily yield an accurate reconstruction of evolutionary history. Different genes may yield different putative phylogenies, for the reasons suggested above. The survey of a broad range of genetic data is necessary to produce a phylogeny in which confidence can be securely placed.

Scoring the Data

By the mid-1990s, scientists had gained sufficient experience with molecular phylogeny to draw a more balanced assessment of its relative power. Gone are the heady days when some molecular biologists believed that molecular phylogenetics might replace or at least eclipse morphological endeavors. Instead, debate is vigorous as to how the two approaches are to be utilized in partnership to pursue a shared goal. Molecular and morphological data still give conflicting phylogenetic conclusions sometimes, but congruence is more common. An outcome of the two decades of sparring between molecular and morphological camps has been a sharpening of efforts on both sides to improve their techniques, through recognizing weaknesses and enhancing strengths. This section will briefly review the advantages of each approach.

Paradoxically, one of the strong points of comparative morphology is the simpler nature of morphological homology, *when it can be reliably recognized*. Also paradoxical is an advantage that stems from the erratic nature of the tempo of morphological evolution. An important feature of evolution is adaptive radiation, which occurs at the establishment of a new group, when diversification yields many lineages with their own unique features that subsequently change little. If such a radiation occurred deep in evolutionary history, a clocklike accumulation of genetic mutations would be unable to track the details of the brief burst of change, for the following reasons. A slow rate of mutation in DNA sequences would leave the event unrecorded. DNA sequences that change rapidly, on the other hand, *would* be able to capture such change; however, this information would be overwritten to the point of illegibility by subsequent mutation. By contrast, the morphological changes that accompany the radiation would, in principle, remain present in the lineages' subsequent history, preserving a record of the event for comparative morphologists to discern. The rapid radiation of eutherian (placental) mammals toward the end of the Cretaceous period, 100 million years ago, is a good example here.

Museum collections represent centuries of effort at cataloging nature, including both fossil specimens and samples of recently extinct species, of which there is an alarmingly large number. These are as accessible to the comparative morphological approach as are living species. Such is not the case with molecular phylogenetics, which works best with DNA from living tissue. Although techniques are being developed to extract and analyze genetic material from dead organisms (see final chapter), comparative morphology is likely to retain an edge here for a very long time, perhaps indefinitely.

A major advantage of molecular phylogenetics is the potential extent of information it can evaluate, which at the limit is equal to the entire genome (in humans, for example, this is three billion nucleotides). Morphological characters necessarily represent only a subset of this information. And because different sectors of the genome accumulate mutations at very different rates, genetic methods offer access both to ancient divergences (by reference to slow-changing DNA, such as ribosomal DNA, which codes for part of the cell's protein-manufacturing machinery) and to recent events (by reference to fast-changing DNA, such as mitochondrial DNA, which codes for components of the cell's energy-yielding apparatus). Morphological information cannot encompass this range of evolutionary history. It is also powerless to discern evolutionary history where physical characters are simple and limited, such as in the early divergence of microorganisms and the origin of chloroplasts and mitochondria. Molecular phylogenetics is able to reach into these histories and has retrieved some interesting surprises, to be described in Chapter 3.

Morphological phylogenetics, with a long history of its own, has accumulated far more conclusions about the history of life than has molecular phylogenetics, which is still in its infancy but growing fast. In 1993, Colin Patterson and two colleagues surveyed the two fields and compared the separate and joint achievements of each frame of reference. "Congruence between molecular phylogenies is as elusive as it is in morphology and as it is between molecules and morphology," they concluded, somewhat pessimistically. "As morphologists with high hopes of molecular systematics, we end this survey with our hopes dampened." Most observers would show more optimism, explaining the negative remarks of Patterson and his colleagues as the result of their dwelling too much on details rather than looking at the larger picture.

Part of that larger picture is a nascent but lively debate about how molecular and morphological data should be handled when addressing the same phylogenetic question. Should they be kept separate, in pursuit of an independent consensus between them that would lend confidence to the conclusion? Or should they be combined, seeking power in the joint effort? Researchers are about equally divided over which offers the better approach.

Checking the Methods

The raw data, whether molecular or morphological, are just the starting point of reconstructing an evolutionary tree. Half a dozen analytical methods have been developed, some of which work with distance data (such as from DNA-DNA hybridization and immunological measures), others with shared-derived characters versus primitive traits (such as protein or DNA sequences). Whatever the method, the task is formidable. Even with just a handful of species, the number of possible evolutionary trees is vast. With a

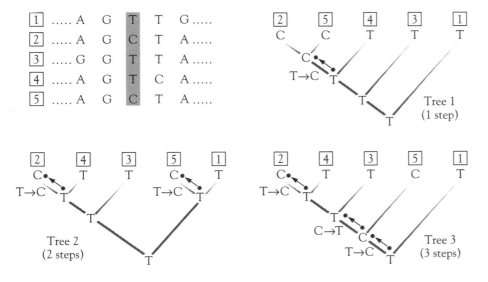

The parsimony technique is the preferred method of deducing phylogenies. Here, we see five individuals and part of their DNA sequences. By concentrating on position three (highlighted at upper left), the parsimony technique seeks to find the tree in which the fewest mutational steps link all individuals. Three such trees are seen here, requiring one, two, and three steps to link the five individuals. The preferred phylogeny is tree 1.

49

mere 50 species, for example, there are 2.8×10^{74} possible trees, which is some 10,000 times as many trees as there are atoms in the universe. A computer that could scrutinize a trillion trees a second (no computer even approaches this speed as yet) would take 8.9×10^{54} years to do the job—that is 2×10^{45} times the age of the Earth. Analytical methods therefore have to negotiate this challenge by rapidly seeking the most likely tree. In reality, many trees are produced, each with equal or nearly equal probability of being correct. Statistical methods are then required to narrow the possibilities.

Methods employing the parsimony principle are currently the most popular and most powerful for phylogenetic analysis. An ancient principle in biology, parsimony looks for the simplest explanation, in the belief that this is the most likely. In the context of phylogenetic analysis, the method looks for the tree (or trees) that utilize the fewest changes to link the given species in an evolutionary hierarchy—the shortest distance between the points.

As mentioned earlier, however, even with the most powerful of methods it is not possible to *know* whether the finally selected tree is the historically correct one. The best conclusion is simply the best hypothesis of what might have happened; short of traveling back into time there is no way of determining whether it is correct. An important sideline activity in phylogenetics, therefore, is finding ways to check the efficacy of analytical methods.

Two have been developed. The first works through numerical simulation to produce artificial phylogenies; the second utilizes known phylogenies of living organisms. Numerical simulation tries to mimic evolution in a computer, using assumptions about evolutionary mechanisms to produce a known evolutionary history. Methods of phylogenetic analysis are applied to data from the end products of this artificial evolution. Their conclusions can then be measured against the known history that the computer had produced earlier. The advantage of this approach is that very many phylogenies can be generated relatively easily. The disadvantage is that even in the most sophisticated models, the evolutionary assumptions are woefully simplistic compared with what is known about molecular evolution. Another drawback is that some of these same assumptions about evolutionary mechanisms (such as rate and mode of mutation) are incorporated into the analytical techniques themselves. As a result, the analytical methods are

A micrograph showing freeze-dried particles of bacteriophage T7 reveals the virus's icosahedral shape and the tail through which it injects DNA into its hosts.

in effect tuned to the way the phylogenies are generated, loading the odds toward a favorable outcome of the supposedly objective test.

The second avenue, that of using the known evolutionary histories of living organisms, is more powerful, because it is based on real evolution—and more difficult to achieve. As David Hillis and his colleagues recently noted, "Experimental phylogenies provide a reality check on simulation studies and provide a test of the fallibility of analysis methods." The work of Fitch and Atchley with inbred mice is an example of the use of known histories; in fact, they embarked on this research explicitly as a way of testing analytical techniques.

Recently another system has been developed, which overcomes some of the problems of the mouse system. It tracks successive generations of the bacteriophage T7. Bacteriophages (a term usually shortened to phages) are viruses that parasitize and destroy bacteria. They have the research advantage of a short generation time—measured in minutes rather than (as with mice) months—therefore allowing more data on which to base realistic phylogenies. It is also easier to obtain extensive genetic information from phages than from mice. This not only helps in the phylogenetic analysis but also gives insights into the details of change at the molecular level, aspects of which can then be incorporated into simulation models. The researcher creates the phylogeny by separating phage colonies at chosen times, eventually producing a series of lineages whose history is precisely known, offering a controlled test of phylogenetic methods.

In a recent survey of these several tests of phylogenetic analysis, Hillis and his colleagues concluded as follows: "Both simulation and experimental phylogenies indicate that many methods are powerful enough to reconstruct evolutionary histories with a high degree of accuracy." This positive view is, of course, important as the application of phylogenetic analysis expands into many fields, subsuming biological comparisons on all scales, from the entire tree of life to simple laboratory investigations such as the T7 phage story.

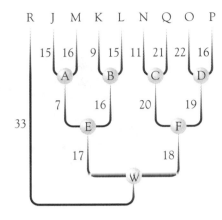

The tree shows a known phylogeny of bacteriophage T7. W represents the ancestral population, which is divided into two subpopulations, E and F. The numbers 17 and 18 represent the number of mutations that accumulated in the two populations after a specified amount of time. E and F were then each divided into two subpopulations, A and B, and C and D. Again, the numbers represent the number of mutations that had accumulated after a specified time. The process is then repeated once again, yielding eight populations in all. The DNA sequences of these eight populations are then analyzed by different types of phylogenetic methods, to produce inferred phylogenies. A comparison of these inferred phylogenies and the known phylogeny gives a test of the efficacy of the different methods.

The Tasmanian wolf, an extinct example of convergent evolution, closely resembled the North American wolf, despite being evolutionarily separate for more than 100 million years.

Trees of Life

*M*olecular phylogenetics may be in its infancy, but its short life has been remarkably productive. In the great majority of cases, its techniques have served to confirm evolutionary histories that had been strongly inferred by means of traditional phylogenetic methods. But in many regions of the tree of life, traditional methods have either failed to yield an unequivocal answer or, because of the insufficiency of morphological information, produced no answer at all. It is in these areas that molecular phylogenetics is potentially of greatest value.

Chapter

3

The most important example here is the phylogenetic relationship among microorganisms, which, if known, would implicitly indicate the universal ancestor of all forms of life. Documenting this relationship requires scrutinizing evolutionary history at its most ancient, and would reveal the earliest branches of the tree of life. Molecular phylogenetics is also particularly powerful at revealing recent evolutionary events, and has even been used to reconstruct the infective trail of HIV (the AIDS virus) from person to person.

Between these two temporal extremes, molecular approaches enjoy fewer advantages over morphological methods. This middle ground of evolutionary history documents the relationships of the major groups, such as the radiation of metazoa (multicelled animals); and the radiations *within* this group, such as the history of the land vertebrates; and the radiations within *these* groups, such as the diversification of mammals; and so on through the hierarchy of the tree of life, going from the most ancient, ancestral forms to the most recent, descendant forms.

Such documentation fills in the details of the tree of life, showing where the principal branches are. Molecular phylogenetics has already contributed new insights into some of the many remaining questions. When the complexities of evolution at the molecular level are understood more fully and methods for analyzing molecular data are further honed, molecular phylogenetics will be poised to yield considerably more progress here, eventually promising to generate a complete picture of the tree of life. For a fraction of the cost of the human genome project (which will approach $2 billion by the year 2005), such a goal could be achieved within a decade. Unfortunately, molecular phylogenetics is not viewed as a glamorous venture by funding agencies. Progress in mapping the tree of life will therefore depend as much, or more, on limitations of resources as on the technical ability to do it.

A note on imagery and terminology may be useful here. As noted, biologists often speak of the tree of life, because the shape of a tree reflects the shape of evolution through the passage of time. It begins with the trunk (the universal ancestor); splits into major branches (the principal forms of life, such as animals, plants, fungi, and so on); and these branches progressively bifurcate, yielding many smaller branches (the groups of organisms within the principal forms, such as mammals, birds, reptiles, and amphibians within the vertebrates); terminal twigs represent the extant species of

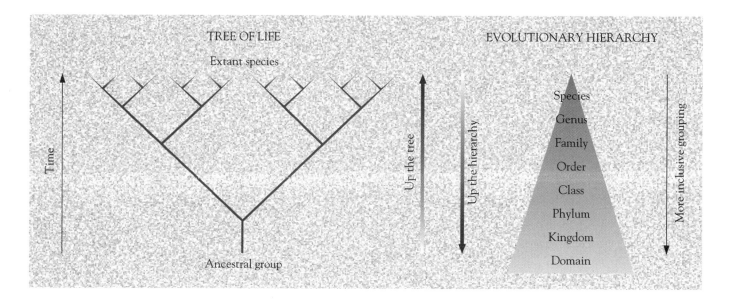

TREE OF LIFE

Extant species

Time

Ancestral group

Up the tree

Up the hierarchy

EVOLUTIONARY HIERARCHY

Species
Genus
Family
Order
Class
Phylum
Kingdom
Domain

More inclusive grouping

all groups. The imagery works well, and for that reason is appealing; but it conflicts with terminology used to describe life's hierarchy. In the tree image, the deepest branches of life, which represent the unfolding of the major groups, are at the bottom; the species, which are the finest level of detail of those groups, are at the top. Going from the twigs to the trunk is, therefore, going *down* the tree. And yet, when biologists speak of life's hierarchy—going from species to genera to orders to families, and so on to the major kingdoms—they speak of going *up* that hierarchy. The confusion flows in part from a conflation of two perspectives. One—the tree of life idea—views the image as representing evolution unfolding through time, the most ancient at the trunk, the most recent in the twigs. The other—life's hierarchy—represents a classification of species within ever-more inclusive groups (genera, orders, and so on). As long as this distinction is remembered, the *up* or *down* orientation should cause no problem.

This chapter will present some examples of molecular phylogenetics. By illustrating where and how effectively it has been applied, its current limitations will also become apparent. The survey will begin at the top of evolutionary hierarchy—that is, the earliest branches of the tree of life—and work down to the relationships within groups or organisms.

By convention, a tree of life is illustrated with the ancestral group at the bottom, going *up* to extant species. By contrast, in the classification of organisms, the domain to species direction is *down* the hierarchy.

Kingdoms Questions

The classification of organisms is a description of the relationship among them in a hierarchical arrangement, with kingdoms at the highest level and species at the lowest. As we saw in the previous chapter, in pre-Darwinian times this hierarchy was considered the product of design. In post-Darwinian times it is considered to reflect evolutionary history, or genealogy. Since the turn of the century, ideas about how the overall structure of life should be classified have changed repeatedly and substantially. The most recent example was the direct result of a totally unexpected discovery in molecular phylogenetics.

Early natural philosophers viewed life as a simple dichotomy: all life forms were either animals or plants. When, three centuries ago, microorganisms were discovered, they were accommodated within this scheme: the large, mobile organisms were denoted as animals, and the smaller, less mobile ones (including bacteria) denoted as plants. Ernst Haeckel, the German paleontologist, challenged this simple dichotomy in the mid-nineteenth century, arguing that the single-celled organisms known as protists (those other than bacteria) fit neither category. Many are photosynthetic, like plants, but have large cells and can move about, like animals. The tree of life therefore came to have three main branches, not two: Animalia, Plantea, and Protista. Early in this century the bacteria were assigned kingdom status (Monera), giving four branches to the tree. The final change made during the premolecular phylogenetic era came in 1959, with the allocation of kingdom status to fungi. Despite certain logical inconsistencies, this five-kingdom scheme, advocated by Robert Whittaker of Cornell University, became firmly established, with minor modification, as the most reasonable description of life's order: Animalia, Plantae, Fungi, Protista, and Monera.

One of the inconsistencies is as follows. The difference between Monera and the four other kingdoms is far greater than that among these four, and yet all have the same taxonomic rank. More logically, the primary division of life is between Monera and the Animalia/Plantae/Protista/Fungi group, with the distinction between animals and plants, for instance, being secondary.

Such a primary division had, in fact, been recognized for more than a century by those who study cell structure. Bacteria are prokaryotes, that is, cells in which the genetic material is not packaged within a nucleus; all

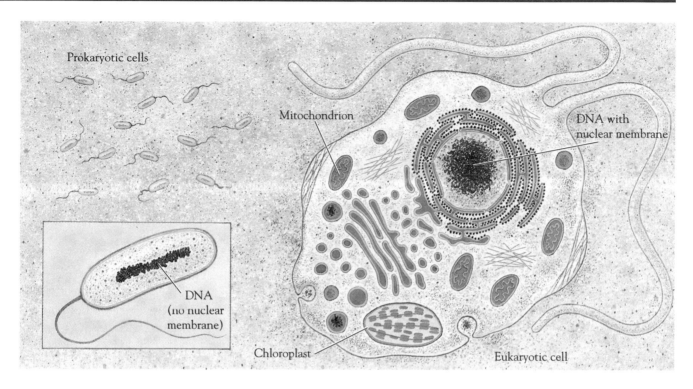

other forms of life are eukaryotes, that is, cells with nuclei. Prokaryotic cells are typically much smaller than eukaryotic cells and contain less genetic material. The prokaryote-eukaryote division, which is based fundamentally on features of cell structure, became assumed to be an evolutionary division, too. This division was formalized as two primary kingdoms, the Procaryotae and the Eucaryotae. The unity of the eukaryotes seemed reasonable, as they shared many complex characters: they could be seen as being descended from a single ancestor. The prokaryotes were also assumed to descend from a single ancestor, despite a deep ignorance of evolutionary relationships within this group. The two-kingdom and five-kingdom schemes coexisted, despite the apparent logical primacy of the former.

Prokaryotic cells are smaller and have a simpler internal structure than eukaryotic cells. In prokaryotes, for example, chromosomes are not packaged within a nuclear membrane. Eukaryotic cells contain mitochondria (the powerhouse of the cell) and, in photosynthetic species, chloroplasts; both these derive from prokaryotes, through an ancient endosymbiotic event.

New Domains Comprise Old Kingdoms

Microbiologists had labored long to elucidate relationships among the prokaryotes, but were hampered by a paucity of physical characters to compare. By the 1960s, most had abandoned the effort and declared the problem unsolvable with traditional methods. Molecular data offered a means of seeking relationships within the group, however, and when rudimentary DNA sequencing methods became available in the late 1960s, Carl Woese and his colleagues at the University of Illinois began to apply them. Since the discovery of ancient microfossils, bacteria were known to have evolved within a billion years of the Earth's formation some 4.5 billion years ago. It was therefore necessary to select a molecule that could record events of such antiquity—a molecule whose function, and therefore its structure, had remained little changed throughout the history of life. The RNA of ribosomes, the structures upon which proteins are manufactured, offered such a possibility. There are three RNA molecules, or subunits, in ribosomes: large, small, and very small (denoted 23S, 16S, and 5S, based on their sedimentation properties during laboratory isolation). Ribosomes are numerous in all cells, and are readily isolated.

Woese and his colleagues compared small sequences of the 16S ribosomal RNA (rRNA) among a large selection of bacteria. They found that most of the bacteria formed a large, coherent group that could be divided into a number of major branches. Some, however, did not fit this group and appeared to form their own, separate group—as ancient as the first, despite being virtually identical in many details of their cell structure. Woese called the first group Eubacteria (true bacteria) and the second Archaebacteria (ancient bacteria). Eubacteria and Archaebacteria are as genetically distinct from each other as they are from eukaryotes, a fact that undermines the supposed primary division between Procaryotae and Eucaryotae. The Procaryotae is composed of two groups, then, the Eubacteria and the Archaebacteria, which are of equal taxonomic status and also equivalent to the Eucaryotae.

In the almost two decades since the Archaebacteria were discovered, the ribosomal RNA of more than a thousand species has been analyzed for its fundamental characteristics, and the complete sequence of the small subunit determined. Other molecules involved in protein manufacture have been scrutinized, too, including the RNA polymerases, which make RNA copies of DNA sequences, and certain factors involved in the translation of genetic information into protein sequence. In all cases, the results have upheld the primary division between eubacteria, archaebacteria, and eukaryotes. Moreover, the evolutionary differences among the three groups are more profound than those that distinguish traditional kingdoms. For instance, all eubacteria have the same subunit pattern in their RNA polymerases, and that pattern is different from that in eukaryotes and archaebacteria. Moreover, eukaryotes are unique in using three RNA polymerase functions. No such differences distinguish, for instance, plants and animals. For these reasons, in 1990 Woese and his colleagues proposed a new formal classification of all organisms.

The three primary groups would have a taxonomic status above that of kingdom, known as domains, and would be called Bacteria (for the eubacteria), Archaea (for the archaebacteria), and Eucarya (for the eukaryotes). Within each domain would be several kingdoms. For the Eucarya the kingdoms are already well established, but some modifications were proposed; for the Bacteria and the Archaea, major branches within these would assume kingdom status. "Our system is an attempt to bring classification into line with recent understanding of phylogeny that stems from molecular studies," explained Woese and his colleagues.

Although the evolutionary reality of the three primary groups, or domains, is now widely accepted, the system of classification that Woese and his colleagues derived from it is still subject to contention. For instance, Ernst Mayr, one of the most prominent evolutionary biologists of this century, criticized the scheme because it does not reflect the world as we see it. He argued that "the difference in structural organization between prokary-

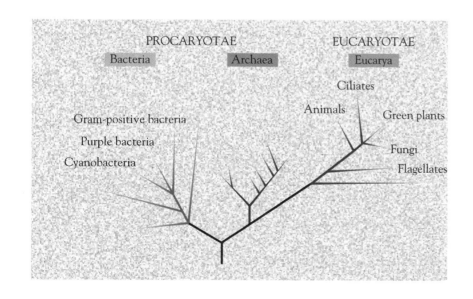

Molecular phylogenetics produced a major surprise when data from ribosomal DNA indicated that prokaryotes comprise two groups, Bacteria and Archaea, whose division is almost as old as life itself. The two groups, or domains, are almost as different from each other as each is from the third major domain, Eucarya, the eukaryotes. This discovery upset the traditional two-kingdom classification of eukaryotes and prokaryotes, which had been assumed to reflect evolutionary history.

otes and eukaryotes is an order of magnitude greater than the relatively small difference between the Archaebacteria and the Eubacteria."

Mayr championed a classification that reflects the traditional division between prokaryotes and eukaryotes. It is unquestionably true that there is greater morphological complexity within the eukaryotes than within the prokaryotes. But it is also true that *at a molecular level* prokaryotes are as complex as eukaryotes, and in their biochemistry as well; in their ecology they are even more complex, as they inhabit a far wider range of habitats. From the perspective of phylogeny, furthermore, the three are equally different from one another. The conflict therefore comes down to one of philosophy of classification. Should it reflect the world as we experience it? Or should it illuminate evolutionary reality?

The Universal Ancestor

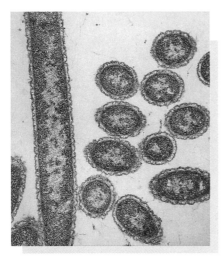

The thermophilic bacterium *Thermus aquaticus* thrives in the hot springs of Yellowstone Park. More rudimentary forms of organisms like this were probably the earliest form of life. The DNA-replicating enzyme from this bacterium is used in the polymerase chain reaction technique because it can withstand high temperatures.

The range of ecological niches occupied by Bacteria and Archaea is both extremely broad and, to eukaryotic eyes, often bizarre. Modern thermophilic (heating-loving) bacteria live in hot springs and the vents of volcanoes, where temperatures are high and oxygen sometimes scarce; others live where levels of salt are high or pressures great. The distribution of these environments among the two domains gives a clue to the conditions under which life arose almost four billion years ago. Common to both domains are thermophilic organisms, which are present in several of the major groups within each domain. Their broad presence implies that thermophiles are the lineages of deepest ancestry. If therefore follows that the earliest organisms, including the universal ancestor of the three primary groups, probably also occupied a harsh, high-temperature environment that sometimes exceeded the boiling point of water—and originated there, too. Most modern eubacteria use oxygen to release energy from carbon compounds, converting it to water and carbon dioxide in the process. Four billion years ago, atmospheric oxygen was scarce, and the earliest organisms utilized other chemicals for this process, such as sulfur, which they converted to hydrogen sulfide. They used the energy released in this process to build organic molecules, using carbon dioxide as the basic raw material.

This description of the universal ancestor, deriving directly from molecular phylogenetics, is fundamentally different from the classical view, which presumed that the first organisms had been heterotrophic (that is,

Columns of fossil stromatolites
exposed in ancient rock in Australia.
Stromatolites were among the
earliest forms of life and comprised
miniecosystems of various forms of
simple microorganisms, going back
more than 3 billion years ago.

had relied on carbon compounds such as glucose to build more complex molecules, like modern metabolic systems do) and lived at temperatures with which we are familiar today. This view was arrived at simply by extrapolating from many modern bacteria that live in temperate rather than extreme environments. Four billion years ago, when life first arose, the planet was still cooling from its fiery birth and there would have been many more environments with extreme physical features than exist today.

The rapidity with which life arose is impressive, as is the rapidity with which it differentiated into the three primary domains. Fossils found in rocks in Australia indicate that photosynthetic eubacteria (and, by inference, archaebacteria and possibly eukaryotes), descendants of the universal ancestor, already existed 3 to 4 billion years ago. This implies that the universal ancestor existed earlier and was more rudimentary than its descendants in its cellular and genetic structure, and was possibly based on RNA rather than DNA.

So far, the three primary domains have been described as being equally related to each other, which leaves the tree of life without a root. Faced with such a situation, phylogeneticists would normally seek an outgroup with which to compare the taxa under consideration. In this case, of

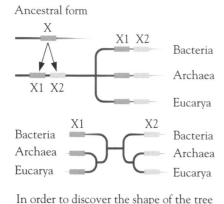

Ancestral form

In order to discover the shape of the tree of life at its deepest roots, it is necessary, as a substitute for outgroup comparison, to exploit paralogous forms of a gene in the universal ancestor that duplicated prior to the divergence of the Bacteria, Archea, and Eucarya (top). The sequence of the paralogous genes X1 in the three lineages is then compared with those of the paralogous genes X2 in the tree lineages. The result (bottom) shows that the sequences in Eucarya and Archea are more similar to each other than either is to those in bacteria.

course, because *all* life forms are under scrutiny, there are no other life forms to refer to as an outgroup.

One way around this is to look at genes that duplicated prior to the divergence of the three primary lineages. Recall that genes that duplicate are termed paralogous, a form of molecular homology within species. In the description of paralogy given in Chapter 2, we imagined that a gene X duplicated in an ancestral species, giving the paralogous genes X1 and X2. The two genes begin accumulating mutations as soon as the duplication has occurred. When the ancestral species later diverges, yielding two lineages, mutations will continue to accumulate in genes X1 and X2 in the two lineages, but only now will genetic differences begin to accumulate between, say, the X1 gene in one lineage and the X1 gene in the other. This means that there is a greater genetic difference between X1 and X2 within one descendant lineage than between the X1 genes of the two descendant lineages and between the X2 genes of these lineages. Similarly, if one of these lineages itself later diverges, genetic differences will begin to accumulate in gene X1 among the two descendant lineages and in gene X2 among the two lineages. The genetic difference between X1 and X2 in these lineages will, however, already be substantial. The order in which the lineages diverged can be inferred by comparing the sequences of paralogous genes within and among all the lineages.

This method has been applied to the relationship among Bacteria, Archaea, and Eucarya, using sequence data acquired from genes that code for proteins known as elongation factors EF-1 and EF-2, which are involved in the transcription of a DNA sequence into a messenger RNA molecule. The result implies a closer relationship between Archaea and Eucarya than either enjoys with Bacteria. It roots the tree with the Bacteria, suggesting that Archaea and Eucarya shared a common ancestor after Bacteria diverged. It also offers insights into the nature and origin of the eukaryotic cell.

It should be noted that certain aspects of the small subunit of ribosomal RNA are more similar between Archaea and Bacteria than between either of the two domains and Eucarya, suggesting that it is Archaea and Bacteria that are more closely related. A relatively slow rate of evolution of ribosomal genes in Archaea could, however, account for this difference. More seriously, the three-domain system is challenged by some researchers, particularly James Lake at the University of California at Los Angeles. Lake suggests that the Archaea is not descended from a common ancestor, and that the extreme thermophile group, or Eocytes, represents the closest relative to the Eucarya, whereas the rest of the Archaea are closer to the Bacteria. Debate continues over these issues.

Nevertheless, the close relationship between eukaryotes and archaebacteria (whether as a whole or in part) seems strongly supported. In this case the origin of eukaryotes should be sought in the archaebacteria. But genomes in archaebacteria are small compared with those in eukaryotes. How is this paradox to be resolved? Two insights may provide clarification. First, many genes in eukaryotic genomes are members of families that have duplicated several times in relatively recent history, so the genome ancestral to modern eukaryotic genomes would have been smaller. The second insight has come from making direct comparisons between genes of eukaryotes and archaebacteria. It has proved possible to imagine what the ancestor of modern eukaryotic genes might have looked like, by determining their most fundamental components. These inferred ancestral genes are simpler in structure than the modern genes. A great deal of similarity is apparent between the structure of these inferred ancestors and their equivalent genes (homologues) in modern archaebacteria. This similarity has been taken to imply a close evolutionary relationship, despite its antiquity. Such comparisons are, however, few so far, so the extent of this apparent ancestral relationship remains to be seen.

This and the work of Woese and others has been taken to imply not only an early origin of life, but also an early origin of all three domains, somewhere close to 3.5 billion years ago. Not all molecular biologists agree. Early in 1996, Russell Doolittle, of the University of California at San Diego, threw down a challenge to what has become accepted wisdom. Doolittle employed protein sequences as a molecular clock, rather than the more commonly used DNA, not least because he was able to survey a large number in 15 groups of organisms, from bacteria to baboons. Assuming that the 57 proteins he studied accumulated change at a regular rate, Doolittle concluded that although life originated some 3.5 billion years ago, the major domains—the eubacteria, archaebacteria, and eukaryotes—did not become established until close to 2 billion years ago. Among the questions raised by this scenario is, What kind of organism existed before the establishment of the three known domains?

Doolittle's interpretations are so at variance with those based on DNA sequences that, not surprisingly, they have attracted strong criticism. Some critics point to the fossil evidence, which shows microfossils that resemble modern bacteria in rocks dating to 3.5 billion years ago. According to Doolittle, such organisms arose only 2 billion years ago. Other critics attack Doolittle's assumption of a constant rate of change in protein structure, even though Doolittle presents a strong argument in its favor. With no resolution in sight of the differences of opinion here, the episode shows that, powerful though molecular phylogeny may be, it is not simple.

Genomes Beyond the Nucleus

Although in eukaryotes the bulk of the genetic material is housed in the nucleus and derives from an archaebacterial ancestor, there are two other genomes in eukaryotic cells—namely, those in mitochondria and in chloroplasts. Are they also of archaeal origin? Mitochondria house metabolic machinery that break down nutrients, such as carbohydrates, thereby releasing and harnessing energy that cells use in their manufacture of molecules of all types, including other carbohydrates, proteins, fats, and nucleic acids. The metabolic processes that release and harness energy consume oxygen and yield carbon dioxide as a by-product. Often called the powerhouses of the cell, mitochondria are present in all but the most primitive eukaryotic cells. Chloroplasts trap the sun's energy in the process of photosynthesis, producing energy-rich molecules that, again, are used in the manufacture of complex macromolecules. Photosynthesis consumes water and carbon dioxide and yields oxygen as a by-product. Both organelles contain genomes that direct protein synthesis; but both depend on genes in the nucleus for much of their structure and metabolic activity, mitochondria more so.

The origin of these organelles in eukaryotic cells has been a matter of discussion ever since their discovery more than a century ago. Two principal theories have been popular. The first (origin from within) argued that both mitochondrial and chloroplast genomes arose when nuclear genes became split off and became packaged within an organelle. In the second (origin from without) the two organelles were viewed as evolutionary remnants of bacteria that had been engulfed by the ancestral eukaryotic cell.

Both organelles have the physical appearance of bacteria, which has been adduced in support of the "origin without," or endosymbiosis, theory. But their genomes are smaller than those of bacteria by at least an order of magnitude, lending support to the "origin within" theory. It proved impos-

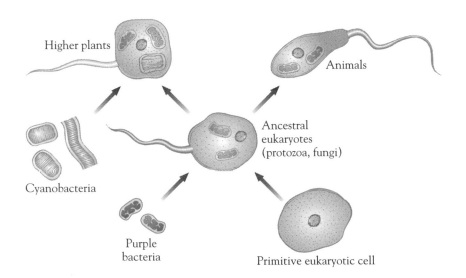

Higher plants

Cyanobacteria

Purple
bacteria

Animals

Ancestral
eukaryotes
(protozoa, fungi)

Primitive eukaryotic cell

The origin-from-without, or endosymbiosis, hypothesis argues that chloroplasts and mitochondria were once free-living prokaryotic organisms that were long ago engulfed by the primitive eukaryotic cell, with cyanobacteria being the precursors of choloroplasts and purple bacteria the precursors of mitochondria. Molecular data strongly support the endosymbiosis hypothesis, although questions still remain about the number of times engulfment occurred, for both chloroplasts and mitochondria.

sible to resolve the issue unequivocably based on physical attributes. Molecular data, particularly that acquired from ribosomal RNA, again offered a way of settling this ancient evolutionary problem. The origin of chloroplasts from bacteria (specifically, the photosynthetic cyanobacteria) was demonstrated more than a decade ago, through comparison of ribosomal RNAs. One question remains, however: Are chloroplasts the descendants of a single common ancestor cell or of multiple ancestors? In other words, did the endosymbiotic union that gave rise to chloroplasts occur just once? Or did it take place many times? Several types of chloroplast exist, usually distinguished by their pigments, so multiple symbiosis apparently occurred. It is unknown, however, whether these repeated events took place within common descendants of the original bacterial lineage or as several independent events in different bacterial lineages.

Resolving the origin of mitochondria proved more difficult, mostly because the molecular biology of these organelles is rather varied, for instance, in terms of ribosomal structure, genome organization, and the molecular machinery of protein synthesis. In many of its details, the molecular biology of mitochondria is unlike that seen in eukaryotes, bacteria, or archaebacteria. There was therefore no simple comparison to be made with the molecular biology of existing bacteria: sometimes similarities could be identified, but often many differences existed, too. Nevertheless, comparison of the DNA sequences of ribosomal RNA genes finally demonstrated the origin of these organelles from purple bacteria. Once again, the question of single or multiple origin remains unresolved.

The small size of mitochondrial and chloroplast genomes, as well as their dependence on nuclear genes for many housekeeping functions, implies that following engulfment by the ancestral eukaryotic cell, the erstwhile bacteria lost or transferred many of their genes to the nucleus. This means that not only do we, and all eukaryotes, harbor bacterial descendants in our cells, but we also have bacterial genes in our cell nuclei, the origin of which is billions of years deep in the past. These insights into modern life in its overall shape and great diversity have been made possible only through techniques of molecular biology and phylogenetics.

Many Cells, Many Questions

Eukaryotes include the most conspicuous organisms in our daily lives, such as animals, plants, and fungi, all of which are multicellular. (The domain also includes a phylogenetically diverse array of microorganisms and seaweeds, the protists or protoctists.) Multicellular organisms include the metazoa or multicellular animals and are classified into more than 30 phyla, groups of organisms that share a basic body plan. The phylum Chordata, for example, encompasses all organisms, including ourselves, that have a spinal cord and a cartilaginous rod, the notochord, extending the length of the body (replaced during development by the vertebral column in vertebrates). For the most part there is general agreement about which organisms belong to which phylum. But there is considerable uncertainty about how the phyla relate to one another in evolutionary terms. As Simon Conway Morris, a paleontologist at Cambridge University, recently put it: "Ideas about metazoan phylogeny are legion and often contradictory."

The reasons for this confusion are not difficult to understand. Phyla have very different body plans; indeed, these differences are the basis of recognition of different phyla. There are therefore few characters with which to ally one phylum with another in evolutionary relationships. Even where features are shared between phyla, problems remain: Do the similarities imply a shared common ancestor? Or are they the result of natural selection shaping similar forms in similar environments, despite an absence of a close evolutionary relationship? It is the old homology-versus-convergent evolution problem, but made even more difficult because of the time depth involved. The trail, as Colin Patterson has observed, is cold, particularly so here. Further, an ignorance of the *direction* of evolution at this time depth makes it difficult to decide whether a shared character is derived or primitive. Yet the ability to make such a distinction is essential if evolutionary inferences are to be made.

Lamenting these formidable problems, which have dogged his profession for more than a century, Conway Morris nevertheless went on to state, "Now the picture has changed forever. The true outline of metazoan phylogeny seems to be emerging." He attributes the change to the impact of

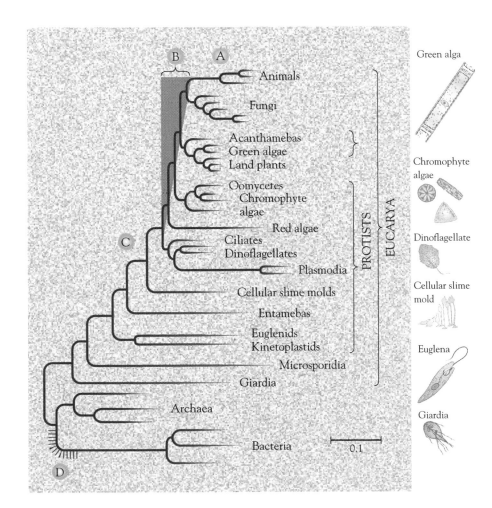

This molecular phylogeny of the eukaryotic organisms was derived from sequence comparisons of the small subunit of ribosomal RNA, analyzed by genetic distance methods. (The scale bar represents 10 changes per 100 nucleotide positions.) The letters indicate major evolutionary events. A is the radiation of coelomate phyla, about 530 million years ago; B, the radiation of eukaryotes, possibly as much as 2 billion years ago; C, the acquisition of mitochondria, possibly as much as 3 billion years ago. D, the root of the tree, is the diversification of the three domains, earlier than 3.5 billion years ago.

molecular phylogeny—particularly, but not exclusively, to the insights gained through comparing ribosomal RNA sequences. Lest this convey too optimistic a picture, it should be stated that more puzzles remain than have been solved—inferences drawn from molecular data often conflict with each other; and there remains the intractable issue that because some important radiations occurred in extremely brief bursts in the distant past, molecular methods may never be able to resolve their branching patterns. As Morris observed, encouraging though the recent progress achieved by molecular phylogenetics may be, it is not yet time simply to "commiserate with the authors of innumerable failed schemes constructed over the past century and promptly move on to more interesting problems."

As one of the three primary lineages of life, the Eucarya has a deep evolutionary history. The lowermost branches of the eukaryotic tree consists of simple microorganisms that have no mitochondria and are confined to oxygen-free environments. Most such organisms live parasitically in the guts of, for instance, cockroaches and termites. The middle branches of the tree are occupied by protists, many of which contain mitochondria but no chloroplasts. The ancestor of later multicellular organisms is somewhere among the protists. An evolutionary burst within these simple organisms produced the major lineages—of animals, plants, and fungi—forming the crown of the eukaryotic tree. A further rapid radiation within the animal, or metazoan, lineage some 530 million years ago quickly established the full range of phyla from which all modern animal species descended, as well as others that subsequently became extinct.

Within this overall shape of the Eucarya, and particularly within the Metazoa, there are many phylogenetic puzzles that traditional methods have failed to solve—and to an understanding of which molecular phylogenetics has contributed. On the large scale, there is the question of the nature of the Metazoa. Is it monophyletic (that is, does it represent all the descendants of a single ancestral form)? What is the relationship of the Metazoa to the other major groups of multicellular organisms, the plants and fungi? How did metazoan body structure—for example, a body cavity in which are suspended digestive and reproductive organs—unfold in evolutionary history? What is the taxonomic status of, for instance, the arthropods? This largest of the animal phyla—the jointed invertebrates—includes insects, crustaceans, and spiders. Is it monophyletic, paraphyletic (that is, containing only some of the descendants of a single ancestor), or polyphyletic (including the descendants of more than one ancestor)? The

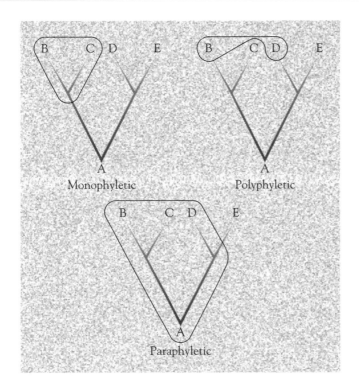

Monophyletic

Polyphyletic

Paraphyletic

A monophyletic group includes all descendant species of a single ancestor, and the ancestral species itself. A paraphyletic group includes only some descendant species of the ancestor. Such a grouping might be thought to reflect nature if, for instance, some species in a group evolve so far away from the ancestral condition at to be treated separately. A polyphyletic group includes species that have converged on a similar adaptation. The latter two reflect nature as adaptation, but do not reflect phylogeny. For some phylogeneticists, such as cladists, the only natural group is a monophyletic group, because such a group is true to phylogeny.

following section will address some of these issues of the history of multicellular organisms, while eschewing the detailed nomenclatural swamps into which it is so easy to sink.

The early application of 18S ribosomal RNA data to the question of the Metazoa as a whole indicated that it included the descendants of more than one ancestor. (The 18S subunit is equivalent to the 16S subunit in Bacteria and Archaea). Rudolf Raff and many colleagues analyzed ribosomal RNA sequences containing more than 1000 nucleotides altogether from 22 species representing 10 animal phyla. They concluded that the Metazoa incorporated two separate groups, the radially symmetrical coelenterates (including sea anemones, corals, jellyfishes, and others) and the Bilateria, or animals with a bilateral body structure. Each derived from a separate protist ancestor. Later phylogenetic analyses of these same data changed this view, however, and the entire Metazoa group was recognized as having descended from a common ancestor, with coelenterates diverging early in the group's history. The flatworms separated next, leaving the

Early phylogenetic analyses of the metazoa indicated that the group as described was polyphyletic; that is, the coelenterates (including sea anemones, corals, jellyfish, and others) derived from one protist ancestor while the Bilateria (animals with a bilateral body structure) derived from a separate protist ancestor. Later analyses, such as this one, showed the metazoa in fact to be monophyletic, with a history like that illustrated here.

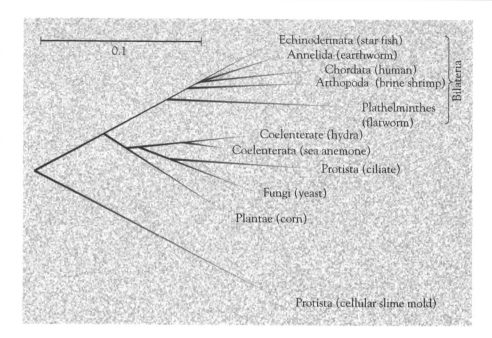

coelomates—animals with a true body cavity—to share a common ancestor before diverging. The coelomates subsequently experienced a radiation yielding three distinct groups, the arthropods, the eucoelomate protostomes (annelids, brachiopods, and molluscs), and the chordates (which includes vertebrates) and echinoderms (which includes sea urchins and starfishes). This pattern now seems secure, following several molecular phylogenetic analyses by different research groups, confirming a view of Metazoan evolution that Haeckel had expressed more than a century ago. In this case, inferences based on morphology had been correct.

Metazoan Relations

The relationship among the major groups of multicellular organisims—the animals or metazoa, plants, and fungi—remains a puzzle for traditional phylogeneticists, despite considerable molecular data that has recently been brought to bear on it. Fewer than five substitutions per thousand positions in ribosomal RNA separate these groups at the crown of the tree of multicelllular organisims, making their phylogenetic resolution extremely challenging.

Several research teams have examined ribosomal RNA sequences and protein gene sequences (including RNA polymerase II genes and others) with mixed results. Three studies indicate that plants diverged first, leaving animals and fungi to share a common ancestor before later diverging. Two studies concluded differently, affirming that fungi diverged first and that animals and plants shared a common ancestor. Analyses of ribosomal RNA and protein-coding genes appear on both sides of this difference of opinion. How are these conflicting results to be interpreted?

The study using data from RNA polymerase II perhaps should be given more weight, for two reasons. First, the 4000 base-pair sequence of the polymerase gene that was analyzed had accumulated four times as many substitutions in the course of evolution than had the 18S ribosomal RNA sequence, thereby yielding more phylogenetically informative data. Second, RNA polymerase II duplicated before the animal-plant-fungi diversification, yielding paralogous genes that allow the shape of the phylogenetic tree to be discerned, as described earlier. One potential problem with this approach is that gene families sometimes evolve in concert, which obscures the evolutionary pattern. Concerted evolution is more likely, however, if the different members of the family perform the same function. Since RNA polymerase genes perform different functions, concerted evolution is less likely to have occurred than in the case of other paralogous genes used to address the animal-plant-fungi relationship. The RNA polymerase study suggests that animals and plants are evolutionarily close, with fungi having diverged earlier. Until more studies are performed, however, this relationship must be regarded as unresolved.

The molecular data support the notion of an explosive radiation of animal phyla 530 million years ago—known in paleontological circles as the Cambrian explosion—rather than a slow, lengthy Precambrian diversification that remains invisible in the fossil record. That explosion involved an irruption of triploblastic phyla—that is, metazoans built from three embryonic cell layers, the endoderm, mesoderm, and ectoderm. The coelome (body cavity) had arrived, thereby permitting animals to become larger and more muscular, allowing them to exploit burrowing niches.

The most species-rich of all the coelomate phyla is the arthropod phylum. There are three major living groups within this phylum: Atelocerata (insects, millipedes, and centipedes), Crustacea (shrimps, lobsters, crabs,

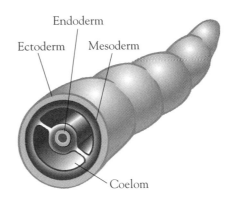

All higher animal phyla develop a hollow space within their bodies, called the coelom, which accommodates the internal organs. The bodies of these phyla are formed from three cell layers, the endoderm, mesoderm, and ectoderm, shown here in a representative embryo. In the formation of the coelom, the mesoderm comes to completely surround the body cavity.

and barnacles), and Chelicerata (horseshoe crabs and spiders). The taxonomic status of arthropods has long been a contentious question in phylogenetics. Morphological and embryological data have been interpreted variously as implying that arthropods are either some of or all the descendants of a single ancestor, although the latter interpretation has been most favored. The molecular data, particularly ribosomal RNA sequences, initially led to even more varied interpretations, including the possibility that the group included descendants of more than one ancestral species. Confusion abounded because of difficulties of aligning sequences, problems arising from rapidly evolving groups, and the obscuring of genetic information by multiple substitutions at single sites. The question now seems to have been settled, however, using a new form of molecular information. Wesley Brown and his group at the University of Michigan, Ann Arbor, turned not to gene *sequences* but to gene *order* in a genome, specifically the order of genes in the mitochondrial genome.

Most metazoan mitochondria have 36 or 37 genes. Two code for ribosomal RNA; 22 for transfer RNA molecules, which are involved in translating genetic information into protein sequences; and 12 or 13 for proteins involved in energy metabolism. The order of these genes on the circular mitochondrial genome differs among different groups. Because the number of gene orders possible with 36 or 37 genes is very large, groups that share the same or similar order probably do so as a result of shared ancestry rather than chance. Comparisons of gene order within the arthropods (with molluscs, annelids, and nematodes as outgroups) strongly indicate descent from a single species, with atelocerates (insects and their relatives) and crustaceans most closely related. The comparison of mitochondrial gene order promises to be a powerful tool for resolving ancient evolutionary relationships.

Vertebrate Variations

The origin and evolutionary relationship of tetrapods (land vertebrates), which evolved 400 million years ago, is both an important and controversial question. Both lobe-finned fish (including extant lungfish) and ray-finned fish have been variously considered the closest relative of tetrapods, with lungfish most favored. The unexpected discovery in 1938 of a coelocanth (*Latimeria chalumnae*), which is related to lungfish in the lobe-finned fish group and was thought to have become extinct 80 million years ago,

The coelocanth, the first specimen of which was found in 1938, belongs to an ancient lineage and was thought to be extinct. It is sometimes called a living fossil.

aroused hope that this organism would provide an anatomical link with the tetrapod ancestor. Comparative morphological analysis of this "living fossil" failed to resolve the question of whether it or lungfish was the sister group of living tetrapods. Five decades after the discovery of the first coelocanth (many more have been caught since), molecular phylogenetics offered another way to address this question.

A remarkable range of molecular data have been examined, including mitochondrial genes, nuclear ribosomal RNA, and structural genes of the nucleus, such as the genes coding for prolactin and growth hormone and genes of the immune system. The most promising line of evidence, surprisingly, was derived from mitochondrial DNA, particularly the slowly evolving sequences of the 12S ribosomal RNA gene and the cytochrome b gene. A range of different trees emerged from these various studies, some of them distinctly unlikely. The mitochondrial DNA favors a tetrapod-lungfish relationship, with the coelocanth as the sister group and ray-finned fishes most distant, although the support for such a tree is not overwhelmingly strong. Once again, the molecular phylogenetic approach is limited by the fact that the evolutionary events under consideration occurred during a 20-million-year window, some 400 million years ago.

Within the tetrapods, one of the most intensely studied groups is the mammals—not least because those doing the studies are themselves mammals. The 3000 living species constitute three groups—the monotremes (egg layers), marsupials, and placental mammals. Reconstructing the evolutionary relationships among placental mammals alone has recently been described by Dan Graur, of Tel Aviv University, as "a formidable phylogenetic task." With at least 15 orders of placental mammals, there are some 10^{12} possible phylogenetic trees linking them, only one of which reflects historical reality. How (to put the question most basically) do carnivores, rodents, primates, insectivores, and so on relate to each other in an evolutionary framework? Many of the most prominent scholars in systematics have tackled this question during the past 150 years, producing schemes that for the greater part are mutually contradictory. The fact that everyone was using the same morphological dataset demonstrates how difficult a task it was—and remains. Once again, a central problem is the small time window (traditionally thought of as being between 65 and 50 million years ago) during which the placental radiation took place.

A common feature of the many evolutionary trees produced was a high degree of "bushiness," with many orders diverging simultaneously. In one recent example, 16 of the 18 orders examined diverged in a five-way split. The bushiness may reflect evolutionary reality, but it seems unlikely, since simultaneous branching is rare in evolution. Alternatively, it could demonstrate that the data under analysis are inadequate to resolve the branching pattern—a more likely conclusion.

Two decades of molecular phylogeny notwithstanding, "higher mammalian phylogeny is still unclear," as Michael Novacek, of the American Museum of Natural History, recently put it. Despite this pessimistic observation, some positive insights have been gained. For instance, each of the three major mammalian groups are confirmed as having descended from a single ancestor, as expected. And molecular and morphological approaches agree in identifying the New World edentates (sloths, armadillos, and anteaters) as an early branch of the tree, perhaps the earliest. With respect to rodents, also an early branch, the two approaches differ, however, a difference that was sharpened in mid-1996 with the publication of complete sequences of the mitochondrial genomes of rats, mice, and guinea pigs. Morphological inferences have designated this group as monophyletic, whereas during the past few years molecular data have increasingly suggested that it is polyphyletic, including the guinea pig (a hystricomorph) and rats and mice (myomorphs) as separate groups with different origins.

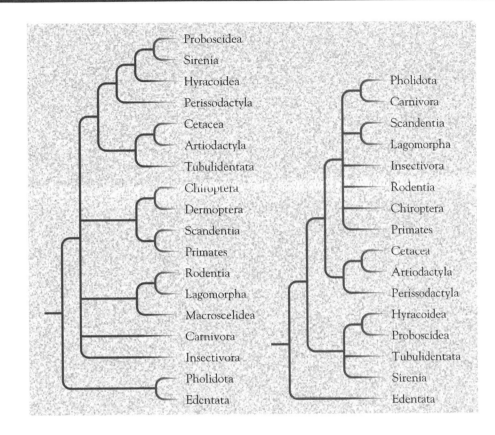

Sorting out the phylogenetic relationships among the mammalian orders is one of evolutionary biology's greatest challenges, because they probably arose rapidly a long time ago. The phylogeny based on morphology (left) essentially divides the orders into two groups, the Edentata and Pholidota, and the rest. Molecular phylogenetics does only a little better, particularly in including a seven-fold split involving the primate order. Inferred multiple divergences of this kind probably reflect a technique's inability to derive more detailed branching rather than evolutionary reality.

The new mitochondrial DNA data, generated by Italian and Swedish researchers, strongly support this (to morphologists) strongly heretical conclusion. Traditional phylogenies have also tended to group rodents and lagomorphs (rabbits and hares) in close association (the Cohort Glires), whereas molecular data deeply undermine this assessment.

A second within-order puzzle that molecular evidence has recently illuminated concerns the Chiroptera, the order made up of microbats and megabats (flying foxes). Most morphologists viewed the order as having a single ancestor, but in the mid-1980s John Pettigrew, an Australian anatomist at the University of Queensland, argued that megabats were closely related to primates, having evolved from a common ancestor, while microbats had evolved separately. The similar wing structure of the two groups of bats was the result, he argued, of convergent evolution. The putative link between megabats and primates was founded on similarities in

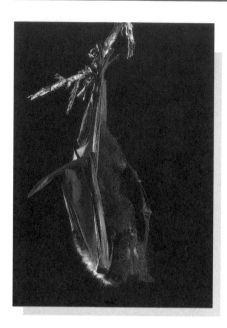

The Indian fruit bat (*Pteropus giganteus*), a megabat. Megabats, or flying foxes, together with microbats, are part of the order Chiroptera. It was recently suggested that megabats might in fact be more closely related to primates than to microbats, based on similar anatomy of the visual apparatus. Molecular data strongly implies that this hypothesis is incorrect.

their visual systems. The neurological structure of certain aspects of the visual system had been assumed to be unique to primates but, as Pettigrew discovered, was also possessed by megabats. Convergent evolution was much less likely to occur in visual neurology than in wing structure, he argued, because of the complexity of the structures involved.

Pettigrew added a second argument against convergent evolution in this case. New brain structures have usually evolved through the modification of existing structures. As a result, new structures incorporate ancestral structural features, even if these features are less than optimal. As the French molecular biologist Francois Jacob once put it, "Evolution works more like a tinker than an engineer." Pettigrew argued that the visual systems of megabats and primates appear to have been assembled on similar ancestral structures, so their similarity reflects common ancestry and is not the superficial result of convergent evolution.

The issue appears to have been settled by Morris Goodman and his group at Wayne State University, who examined a 1200-nucleotide region of the epsilon-globin gene in 17 species. Parsimony analysis of the data strongly allies the two bat groups, to the exclusion of primates. The megabat and microbat uniquely share 39 derived nucleotide sequence changes; megabats and primates share only 3. It appears, therefore, that the similar anatomy of the visual systems in megabats and primates is indeed the outcome of convergent evolution, unlikely though it may seem.

There is certainly more unknown than known in mammalian phylogeny. As Graur recently observed, "Mammalogists are in for some big surprises as far as the higher classification of mammals is concerned."

Graur's comment could not have been more pertinent, because just as this book was going to press, a molecular phylogenetic analysis produced a picture of mammalian evolution that strongly challenges traditional notions. As mentioned earlier, although mammals had coexisted with dinosaurs for 100 million years, they did not undergo an evolutionary diversification until after the terrible lizards had become extinct, 65 million years ago. The sequence of events suggested a plausible explanation, namely, that the absence of the dinosaurs opened up many previously occupied ecological niches, which the basal mammalian stock filled in a brief burst of evolutionary innovation. The new molecular analyses, published in May 1996, indicate that this might not be correct.

Blair Hedges and his colleagues at Pennslyvania State University expoited the fast-growing database of gene sequences to probe mammalian history. They examined sequences of 79 genes in 16 orders of mammals, looking for those that exhibited sufficient clocklike behavior to be reliable for inferring a time of divergence of the orders. About half the genes fulfilled the requirements, and the information encrypted in them produced a surprise. Instead of having diverged in the time window of 65 to 50 million years ago, as indicated by the fossil record, the event apparently took place as much as 100 million years ago and possibly earlier still.

If this conclusion is correct—and other researchers will certainly check it carefully before accepting it—Hedges has an explantion of why the diversification took place when it apparently did. The engine driving the mammalian diversification, suggests Hedges, was not ecological opportunity caused by the sudden absence of dominant competitors, but fragmentation of populations caused by continental drift.

The earliest mammals evolved some 200 million years ago, when the world's continents had coalesced as the supercontinent, Pangea. Mammals, like the dinosaurs, could therefore occupy virtually the entire landmass. By 100 million years ago, the movement of the tectonic plates that compose the Earth's crust had fragmented Pangea into many island landmasses, even more than we see in a map of the continents today. Isolated on different islands as they were, different populations of mammalian species would have evolved in many different directions, just as modern evolutionary theory predicts. During the past 100 million years, many of the previously isolated islands have coalesced, producing the modern continental pattern and allowing the previously isolated species to mingle once again. The geographical isolation described here has earlier been suggested as the engine of diversification of the dinosaurs. Hedges argues that the same thing happened to the mammals. When the dinosaurs became extinct, a great evolutionary diversification of mammals did indeed occur, he concedes, but this was a production of variation on themes that had originated far earlier and was not the origination of those themes, as has been supposed. Hedges accounts for the relative absence of mammalian fossils earlier than 65 million years ago by saying that populations of mammalian species were small, as was the body size of the creatures themselves. This would produce a bias against visibility in the fossil record.

Hedges's work provoked a mixed reception among mammalogists, as is typical when conventional ideas are challenged. Time will tell whether molecular phylogenetics has produced a major insight into one of the more interesting evolutionary events in recent prehistory.

Sorting Out Songbirds

One of the most ambitious forays into molecular phylogenetics investigated the evolutionary relationships of birds, of which there are more than 9000 living species, all descendants of a reptilian ancestor dating to some 150 million years ago. Because of the popularity of amateur as well as professional ornithology, birds are probably the most closely studied animal group of all, and their classification reflects riotous variety. There have been unresolved problems, to be sure, such as the flamingo example given earlier. But ornithologists feel—rightly—much more secure than, say, mammalogists in their overall phylogenetic scheme. Nevertheless, when Charles Sibley and Jon Ahlquist produced their version of the bird tree in the mid-1980s, based on DNA-DNA hybridization, they ruffled many feathers by pointing out some glaring mistakes of classification. Sibley and Ahlquist received some criticism over the reliability of their experimental procedure. And some observers suggested that some of the proposed evolutionary trees could not be supported by Sibley and Ahlquist's data, because inferred genetic distances between species in the tree were too small. For the most part, however, the major revisions proposed by Sibley and Ahlquist have gone unchallenged.

Consider, for example, Australia, land of marsupial mammals, where evolution has produced some unique forms (such as kangaroos) and some forms strikingly similar to Old World species (such as the Tasmanian wolf and the bandicoot, a marsupial that resembles the placental rabbit). Despite this rampant convergent evolution among the marsupial mammals, there was little danger of erroneously inferring a specific evolutionary link between the Australian species and their placental lookalikes in the Old World, because marsupials are clearly distinguished morphologically by the possession of a marsupium, or pouch. Molecular phylogenetics has confirmed what was already known to morphologists.

What of songbirds, however, which make up more than half the world's total of bird species? More than 700 are native to Australia. There are Australian warblers and flycatchers, creepers and babblers, and nectar feeders—all appearing virtually indistinguishable from species known in Africa, Europe, and the Americas. When the task of classifying the Australian songbirds began, in the last century, European ornithologists had already classified birds in most of the rest of the world. The Australian songbirds looked so like familiar species of the Old and New Worlds that it seemed natural to slot them into known groups. Sibley and Ahlquist

The bandicoot, a native marsupial of Australia, resembles the placental rabbit.

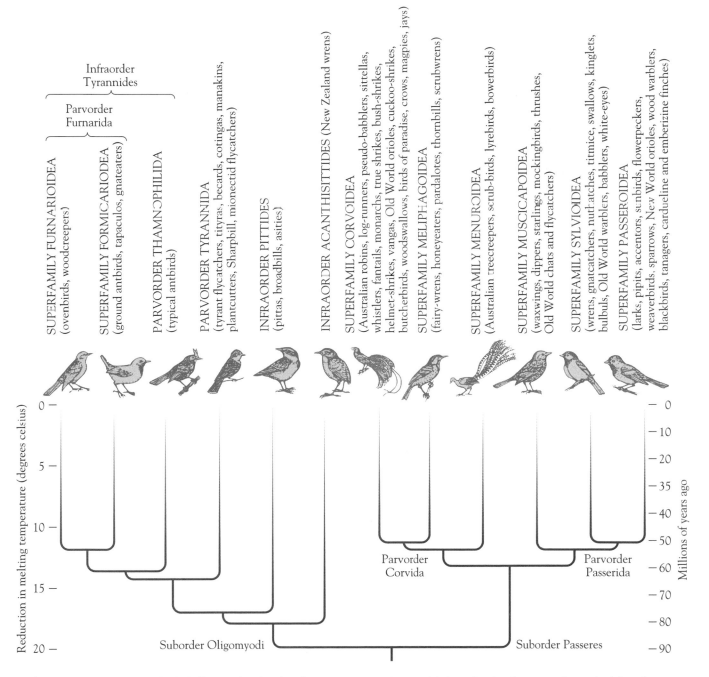

Infraorder
Tyrannides

Parvorder
Furnarida

Reduction in melting temperature (degrees celsius)

SUPERFAMILY FURNARIOIDEA
(ovenbirds, woodcreepers)

SUPERFAMILY FORMICARIOIDEA
(ground antbirds, tapaculos, gnateaters)

PARVORDER THAMNOPHILIDA
(typical antbirds)

PARVORDER TYRANNIDA
(tyrant flycatchers, tityras, becards, cotingas, manakins,
plantcutters, Sharpbill, mionectid flycatchers)

INFRAORDER PITTIDES
(pittas, broadbills, asities)

INFRAORDER ACANTHISITTIDES (New Zealand wrens)

SUPERFAMILY CORVOIDEA
(Australian robins, log-runners, pseudo-babblers, sittellas,
whistlers, fantails, monarchs, true shrikes, bush-shrikes,
helmet-shrikes, vangas, Old World orioles, cuckoo-shrikes,
butcherbirds, woodswallows, birds of paradise, crows, magpies, jays)

SUPERFAMILY MELIPHAGOIDEA
(fairy-wrens, honeyeaters, pardalotes, thornbills, scrubwrens)

SUPERFAMILY MENUROIDEA
(Australian treecreepers, scrub-birds, lyrebirds, bowerbirds)

SUPERFAMILY MUSCICAPOIDEA
(waxwings, dippers, starlings, mockingbirds, thrushes,
Old World chats and flycatchers)

SUPERFAMILY SYLVIOIDEA
(wrens, gnatcatchers, nuthatches, titmice, swallows, kinglets,
bulbuls, Old World warblers, babblers, white-eyes)

SUPERFAMILY PASSEROIDEA
(larks, pipits, accentors, sunbirds, flowerpeckers,
weaverbirds, sparrows, New World orioles, wood warblers,
blackbirds, tanagers, cardueline and emberizine finches)

0 —
5 —
10 —
15 —
20 —

Parvorder
Corvida

Parvorder
Passerida

Suborder Oligomyodi

Suborder Passeres

Millions of years ago

— 0
— 10
— 20
— 35
— 40
— 50
— 60
— 70
— 80
— 90

Birds are among the most studied of all animals. The classification of songbirds had been worked out in some detail before the continent of Australia had been settled by Europeans. When ornithology began to be practiced there, the native birds were grouped with the American and European species, which they strongly resembled. Molecular data recently implied that these resemblances are the result of convergent evolution, and that most Australian songbirds are native to the continent, having evolved independently there. This phylogeny is based on DNA-DNA hybridization.

The Australian treecreeper, superficially similar to creepers elsewhere in the world, is, according to genetic data, a locally evolved species. The resemblance to other creepers is the result of convergent evolution, not recent shared ancestry.

note that "many of the convergences are so subtle that the true relationships . . . probably could not have been resolved from anatomical comparisons alone." Therefore the Australian warblers were assigned to the Sylviidae (true warblers); the Australian flycatchers to the Muscicapidae (Afro-Eurasia flycatchers); the treecreepers to the Certhiidae (Eurasian-American creepers); the honeyeaters to the nectar-eating Afro-Asian sunbirds; and so on. The classification made sense morphologically, although not geographically, since Australia has been separated from the rest of the Old World for at least 30 million years. Only multiple waves of migration could explain this distribution of Australian species. Though not impossible, such repeated migrations from other continents were unlikely. Moreover, the pattern of species' distribution implied by the migration hypothesis is very different from that of the native marsupial mammals, which are exclusively Australian. A geographical concordance of different groups of species is expected, particularly if evolution is local.

The DNA-DNA hybridization data completely undermined the accepted bird classification. It showed the Australian songbirds to be of native Australian origin, bringing the biogeography of the birds in line with that of the mammals. Like the marsupial mammals, the songbirds had become adapted to niches similar to those in the rest of the world and, through convergent evolution, came to resemble their counterparts in the rest of the world, often in astonishingly fine detail. Sibley and Ahlquist suggested that the songbirds of Africa, Eurasia, and North America should be grouped under the suborder Passerida. Australian songbirds, however, would be members of the suborder Corvida. This suborder of songbirds originating in Australia also includes crows, magpies, and jays, which have migrated through the Old and New Worlds.

Lichens, Lakes, and HIV

This final section will offer just a few examples of how molecular phylogenetics has been used to address evolutionary problems at levels in the tree of life below that of class (for example, mammals and birds). Molecular phylogenetics is particularly effective at revealing the recent evolutionary events at these lower levels, including the relationship among species or population histories within species, for which morphological data may be limited.

Lichens are symbiotic relationships between fungi and algae. Usually recognized as feathery grey structures on tree trunks and walls, they in fact can take many other forms. Some fungi-algae symbioses are beneficial mutualisms, while others are harmfully parasitic. Lichens represent some 20 percent of all fungi, but despite years of effort, their place in the phylogenetic tree of fungi has remained enigmatic. They were generally assumed to form a natural group, that is, one stemming from a common ancestor, united by different expressions of the same basic life style.

Since morphological features have proved of little help in discerning the position of lichen-forming fungi on the evolutionary tree of fungi, researchers have recently turned to molecular evidence. For instance, a team at the Smithsonian Institution has examined a sequence of 1927 nucleotides of the small subunit of ribosomal RNA in 75 fungal species, including 10 lichen-forming fungi. Although this survey does not yield a comprehensive picture of fungal phylogeny, it is sufficiently broad to give a first glimpse. The result is very clear. Lichen-forming fungi do *not* form a natural phylogenetic group, despite their clear status as a natural ecological group. Instead, the lichen-forming fungi appear in many branches of the tree, indicating that this life style has arisen several times independently from different ancestral organisms. Lichen-forming fungi appear in the same groups as brewer's yeast, morel mushrooms, and fungal forms that cause plant diseases (such as Dutch elm disease) and in the group that includes *Pneumocistis*, the microorganism that causes fatal pneumonia in AIDS patients.

Furthermore, this phylogeny undermines a cherished idea in ecology—that there is a normal evolutionary progression from parasite to beneficial mutualist. The first stage in such a relationship, it was imagined, would inevitably be parasitic, just as predators and prey are a natural part of life. Unlike most predator-prey relationships, however, that between parasite and host is one of continual physical contact. This being the case, the two species are able to evolve as one, often developing a less aggressive, less one-way relationship, in which both partners benefit. Such mutualism is therefore regarded as the end point of evolution of two species that once were parasite and host. In lichens, however, parasitic and beneficial organisms appear both early and late in their evolution, showing that there is no such progression. It is still early to establish any overall picture of lichenization, but this study has already demonstrated that traditional views (based on limited morphological data) were quite wrong.

The cichlid fish of the African lakes—Victoria, Malawi, and Tanganyika—are both a delight and a puzzle to biologists, not least because of their numbers, totaling more than 500 species. The delight is inspired by the many and often bizarre ecological adaptations they display. Some are algae grazers; others feed on plankton and detritus; while still others prey exclusively on insects or other fish, including their cousins. One species feeds by engulfing the snout of a mouth-brooding female, forcing it to surrender the brood; another feeds on the scales of other cichlids, by using a specially designed jaw and teeth to rasp them; yet another eats eyes, which its jaw is designed to pluck from other fish. The puzzle is over the cichlids' evolutionary origin.

Each lake has a large assortment of these remarkable species (Lake Victoria has 200, for instance), and many display the same or similar adaptations as species in other lakes. And yet the lakes are relatively young, just a million years old in Victoria's case. Two questions must be answered: How did such diversity arise? And how is the pattern of distribution of the different species among the lakes to be explained?

Traditionally, the entire flock of species was classified as a single genus, but recently a cladistic analysis of morphological characters divided them

A species of cichlid fish, *Pseudotropheus tropheops,* from Lake Malawi. Cichlids have apparently undergone a recent and very rich burst of speciation, as indicated by molecular data. More than 500 species thrive in several large African lakes, and, until recently, their evolutionary history was a puzzle.

into a large number of genera, many of which inhabit all three lakes. According to this scheme, at least part of the *within-lake* diversity is therefore explained by an early diversification followed by repeated invasions into each lake. The species flocks in each lake would have derived from many different ancestral species.

In 1990, the first detailed molecular data on the cichlids became available, yielding a very different picture. Allan Wilson and his colleagues sequenced 803 nucleotide positions in mitochondrial DNA, including parts of the cytochrome b gene, two transfer RNA genes, and a region that regulates gene expression; the cichlids studied included 14 species from Lake Victoria and 23 from elsewhere in Africa. The data were quite clear, and quite contradictory of the cladistic analysis of morphology. The range of nucleotide variation among the Lake Victoria cichlid species was tiny, with an average of three differences between the 14 species. By contrast, cichlids from Lake Malawi differed by an average of 50 mutations from the Lake Victoria species. These observations strongly imply that each lake's cichlid species evolved in situ and extremely rapidly, perhaps within the past 200,000 years. This conclusion has since been confirmed by more extensive molecular data and analysis.

The cichlid fishes therefore are a prime example of the uncoupling of genetic and morphological change. The genetic variation among the 200 Lake Victoria species is less than that within, say, a single species of horseshoe crab; and yet the morphological variation among them is extensive. This paradox prompts several questions: What drove such great morphological change in the face of so little genetic change? Further, what was the mode of speciation? Did it take place when each lake was as extensive as it is now, so that new species emerged side by side (that is, sympatrically)? Or did it happen when lake levels were lower, dividing each lake basin into isolated habitats (that is, allopatrically)? So far, the questions are unresolved. The cichlids provide a good example of how phylogenetic insights can inspire wider biological questions.

Finally, phylogenetic analysis helped resolve the question of whether an infected dentist had transmitted the human immunodeficiency virus (HIV) to his patients. In 1990, workers at the Centers for Disease Control, Atlanta, reported the probable infection of a young woman in Florida with HIV from her dentist, who had died of AIDS. After a public announcement encouraged his former patients to submit to an HIV test, a total of 10 were determined to be infected. Interviews with the HIV-positive individuals revealed that 4 of the 10 displayed behavioral risk factors that could have led to their infection by sexual or intravenous transmission. But 6 patients had no risk factors.

HIV mutates extremely rapidly, much more so than most viruses, for reasons that are not fully understood. One medical consequence of this is that it is extremely difficult to produce an effective vaccine. Many vaccines work by stimulating the immune system to recognize the molecular structure of the pathogen's outer surface. If this structure changes constantly, a potential vaccine may be obsolete before it can be used—as with HIV. The virus's rapid rate of mutation also had consequences for the investigation into whether or not the six patients had been infected by the dentist, as suspected.

The first consequence was that each infected individual's virus population was likely to be genetically different, offering the CDC the possibility of making a phylogenetic analysis of the viruses in the six patients and their dentist, which might reveal the origin of the infection. The logic here was as follows. If the virus in question did not mutate or mutated only slowly, then the strains in the patients would be the same as the strain in the dentist. However, this resemblance of strains would not implicate the dentist in the mini-epidemic, because the molecular profile of his virus would be the same as anyone else's. By contrast, a virus that mutates leaves a molecular trail that can be followed: the viruses in a group of individuals collectively infected by person A will be more like A's virus than like B's or C's, and soon. The question the CDC workers asked, therefore, was this: Did the HIV in the six patients descend from the dentist's virus?

The researchers sequenced a gene that coded for one of the protein components of the virus's outer surface, and indeed found genetic differences among the six individuals and their dentist. Phylogenetic analysis of these data indeed revealed that virus strains in the six individuals with no risk factors for HIV infection were evolutionarily close to one another and descendants of the dentist's virus. A series of simulations reconstructing the evolution of the virus demonstrated that, because of the rapidity of HIV mutation, an accurate phylogenetic tree could probably not have been recovered from the molecular data had they been collected more than five years after infection (three years had in fact elapsed). So great a number of genetic changes would have accumulated after five years had passed as to obscure phylogenetic relationships among strains.

This episode represents molecular phylogeny on the shortest of time scales, and serves to emphasize one of the science's greatest contributions—the enormous time range over which it may be applied, from deep evolutionary events some 4 billion years ago to a period of just a few years in the recent past. Phylogenetics as a whole has benefitted from the introduction of a molecular dimension, not least because the perceived competition of a new approach stirred morphologists to enhance the strengths of their traditional methods. Molecular phylogenetics may not be the panacea it was imagined to be in its early, heady days, but it is unquestionably a powerful tool in its own right and, increasingly, in combination with traditional methods.

Mice, *apparently studying the printout of their gene sequences, at Harvard Medical School.*

The Puzzle of
Genetic Variation

We've seen that when biologists wish to reconstruct evolutionary histories of species, they can look at genetic similarities and differences between the species in question. Now, although each species is a cohesive genetic entity, there is genetic variation among the individual members of each species. Some of us have blue eyes, some brown. Some of us are tall, some short. These characteristics are inherited from parent to child, as are facial features: it is usually easy to identify members of different families, for instance.

Genetic variation within a species is the raw material of evolution by natural selection. Some variants may confer adaptive advantages on the

Chapter

4

individuals that possess them. Such individuals will be fitter in a Darwinian sense, and will leave more offspring than those with the inferior variant. In a predator species, for example, individuals that are more efficient at killing prey will be at a selective advantage. Similarly, individuals of a prey species that more effectively avoid predation will be favored by natural selection. Over time, such advantageous features will come to predominate in the species, as will the genes that underlie them.

Most species exist as many geographically separate populations, of course, not just as a single population. As a result, individual populations of the same species often develop distinctive genetic variants. Such variants spread to other populations only when individuals move between populations and begin to mate there. If the populations remain completely isolated from each other, perhaps separated by a major river system or a mountain range, then substantial genetic distinctiveness can accumulate within the populations, sometimes resulting in the establishment of subspecies or even distinct species.

Genetic variation therefore exists between populations of the same species as well as within single populations of a species. We will see in later chapters how biologists are able to exploit this between-population genetic variation in exploring a species' recent history. The question of how and where modern humans evolved is a good example here, as is the long-unresolved issue of when and how humans first entered the Americas. In both these cases, the newly developed genetic techniques have supplemented traditional approaches of anthropology and archeology, producing some surprising results. Genetic variation between populations of individual species has also been used to gain insights into the climatic and geographical barriers that separate them; some of these barriers are historical and are not immediately obvious to present view. We will see in Chapter 6 how an analysis of genetic variation has thrown light on the distribution of many fish, invertebrate, and bird species in the southeastern United States, for example.

The question of the *extent* of genetic variation within and between populations was the subject of long and heated debate among population geneticists in the early decades of this century. Two widely divergent schools of thought emerged. One school argued for the existence of little variation, the other for a lot. Variation could be discerned between individuals at the phenotypic level—that is, in their morphology, as mentioned earlier—but there was uncertainty about how this related to differences in the genotypes—that is, in the package of genetic information, or genome, and its initial products, proteins. Not until the mid-1960s could this issue be addressed directly, when techniques were developed to measure genetic variation at the level of certain differences in protein structure. Techniques developed more recently addressed the same question at the level of the DNA sequence of genes. The question of the *origin* of genetic variation was also closely scrutinized, and the ensuing discussion evolved into what came to be known as the neutralist-selectionist debate. Specifically, is the origin of variation driven by natural selection, which selects favorable new variants? Or are the variants selectively neutral, so that their accumulation is driven by the mutation process itself, and their persistence in a population determined by chance?

Entangled with these arguments was the issue of the *tempo* of the accumulation of variation. Is it regular or erratic? If the pace of the accumulation of genetic variation turned out to be regular, it would offer what has been called a "molecular evolutionary clock." Such a clock would be extremely valuable in pinpointing the time in prehistory when important evolutionary events took place. Three decades since the notion was proposed, the existence of a clock remains a matter of controversy. The concept of the molecular evolutionary clock is the subject of the following chapter.

This chapter will explore two aspects of genetic variation: its extent and the nature of its origin.

How Much Genetic Variation?

Since its establishment as a discipline early in this century, population genetics has had as its goal the description and explanation of genetic variation within and between populations. The goal is the same today, but with a difference. Dating from a technical advance in the mid-1960s, population geneticists have been able to measure genetic variation *directly*. Prior to that advance, geneticists looking for evidence of variation in the genome had to rely on two kinds of information, both of which were *indirect* measures of that variation.

One everyday example of a variation based on genetic differences is eye color. Different variants of the gene that codes for iris pigment may give

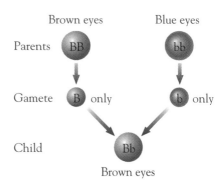

Brown eyes   Blue eyes

Parents   BB   bb

Gamete   B only   b only

Child   Bb

Brown eyes

The allele for blue eyes is recessive to the brown-eye allele. In this case, the offspring of parents, one of whom has two alleles for blue eyes and the other two alleles for brown, will have brown eyes. The offspring is said to be heterozygous for eye color—it has alleles for both colors.

blue eyes or brown. Such variants are called alleles of the gene, and for every gene we inherit two alleles, one from one parent and one from the other. In fact, every heritable character is the outcome of such a pair of alleles, each of which may be dominant or recessive. For instance, brown eyes can result from either a pair of "brown" alleles or one brown and one blue, because the brown allele is dominant and the blue recessive; for eyes to be blue, therefore, both alleles must be the blue variant. These, and other, alleles appear with different frequencies in different populations: blue eyes are common among northern Europeans, for example, but rare among East Asians. Researchers have studied such variation by observing the variants in such popular systems as human blood groups, shell markings in snails, wing patterns in butterflies, and rare recessive mutations in *Drosophila* (fruit flies). Although these systems provided indirect evidence of variation at the level of the gene (or genes) of interest, they gave no basis for predicting how general such variation would be in the rest of the genome. Neither, in most cases, could researchers identify any direct advantage conferred by different alleles that might cause them to be favored by natural selection. For instance, different shell markings in snails confer no obvious benefit or cost.

A second indication of the existence of variation—and in this case its general benefits—came from breeding experiments with, for example, *Drosophila* and corn, particularly with inbred strains, in which genetic variation is generally lower than in normal strains. Breedings *within* inbred strains produced individuals of lower viability and fertility than breedings *between* such strains. The phenomenon of greater fitness from outbreeding implied not only the existence of genetic variation between strains but also suggested that enhancing such variation provides a fitness advantage. Nevertheless, it was not possible to determine whether the effect was the result of variants at a few or many genetic loci. The extent of genetic variation *within a species* therefore remained beyond geneticists' collective grasp. Naturally, these techniques could provide no information about genetic differences *between species*. "Under these circumstances," the Harvard geneticist Richard Lewontin observed recently, "it is not surprising that evolutionary geneticists were divided into opposing schools with more or less uncompromising views of the truth."

The two schools of thought were strikingly different. The "balance" school of Theodosius Dobzhansky and his followers held variation to be extensive, while the classical school of H. J. Muller and his followers held it to be limited. The difference flowed from different assessments of the effect of variation on fitness.

According to the balance school, a population with extensive genetic variation can adapt more easily to local differences in habitat. Should a drought strike, for example, some individuals may have a variant gene that lets them survive with less water, ensuring the survival of at least some of

Theodosius Dobzhansky, a major figure in the development of modern genetic theory, was a leading exponent of the "balance" school of population genetics, which held that genetic variation in populations should be extensive, as the result of natural selection.

H. J. Muller was a leader of the "classical" school of population genetics, which held that genetic variation in a population constituted a burden (genetic load) that would be rapidly removed by natural selection.

the population. Deleterious mutations may occur, but these were held to be soon weeded out by natural selection, and not to contribute significantly to overall variation. Natural selection was therefore seen as promoting variation. Variation between populations was expected to exist, but it was not held to be significant, because variation *within* populations was also high. There was much speculation as to what proportion of genes in a genome might come in different forms, and how many forms of a particular gene might exist. A gene that has such variants is said to be polymorphic, and a gene's level of polymorphism may be high (many variants) or low (few variants). (Genes that have no variants are described as being monomorphic.) Bruce Wallace, a leading proponent of the balance school, speculated that perhaps all genes might be polymorphic.

By contrast, the classical school viewed the role of natural selection as principally to remove deleterious mutations. Only occasionally did a rare, advantageous mutation quickly spread through the population, completely replacing the original allele, producing an optimal genetic package, or genotype. If individuals deviated from this genotype, they would be less fit. High levels of polymorphism were therefore seen as constituting a burden, or genetic load, that reduced a population's fitness. Unless natural selection eliminated this genetic load, a species might become extinct. Variation within a population was therefore likely to be low. Because variation was considered to be limited *within* populations, genetic differences *between* populations was judged to be highly significant. Muller estimated that perhaps as few as one gene loci in a thousand was polymorphic.

Technology Intervenes

The gap between the two schools' philosophies and speculations was great. As Lewontin noted, "Population genetics seemed doomed to a perpetual struggle between alternative interpretations of great masses of inevitably ambiguous data." The theoretical impasse was broken independently in two laboratories by technological innovation—that is, the adaptation of a technique for distinguishing soluble proteins to the study of many proteins in large populations. Examples of soluble proteins include many blood proteins, such as albumin, and many enzymes. An example of an insoluble protein is collagen, which performs a structural role in cartilage, bone, and cell infrastructure. Soluble proteins are necessary for the technique, called electrophoresis, to work, because they must be able to become incorporated into the gel layer on which they are displayed.

Many soluble proteins bear an electrical charge (positive or negative). Under the influence of an electrical current passed through a gel, proteins

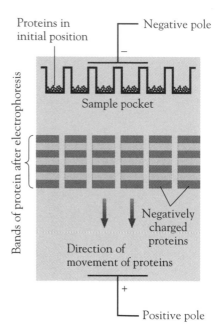

Proteins in initial position

Negative pole

Sample pocket

Bands of protein after electrophoresis

Negatively charged proteins

Direction of movement of proteins

Positive pole

The technique of gel electrophoresis, which separates proteins according to their electrical charge, provided a way of determining the degree of genetic variation in natural populations.

suspended within the gel can be separated (and identified by chemical staining) according to the charge they bear. Proteins with a positive charge will move one way, those with a negative charge will move in the opposite direction; and the greater the charge, the further the protein will migrate from its original position. (A protein's size also influences its movement.) Some amino acid replacements in a protein chain (arising through mutation of the DNA sequence) result in a modification of the electrical charge, allowing such variants to be detected. (The majority of all amino acid replacements do *not* affect the protein's charge, however, and these are invisible to electrophoresis.)

Lewontin, then at the University of Chicago, and his colleague J. L. Hubby developed the technique and applied it to 18 proteins in five natural populations of *Drosophila*. Thirty percent of the proteins were found to exist as more than one variant in the population as a whole. That is, 30 percent of the population's genes that were available for study were polymorphic. Naturally, each individual did not possess the full range of these polymorphisms: even when a gene existed as several variants in the general population, an individual's two copies of the gene were likely to be identical. On average, the two copies were different variants in a little more than one in 10 genetic loci in any individual (the individual is said to be "heterozygous" for such genes). The average percentage of heterozygous genes is therefore said to be, in this case, 11.5 percent. The British geneticist Henry Harris did a similar survey on 10 proteins in human populations. His results were very similar: 30 percent of the proteins were polymorphic, with individuals being heterozygous for 9.9 percent of their genes on average. Lewontin described the extent of the polymorphism he and Harris found as "startling."

The results of these two surveys were published in 1966. They ignited an explosion of similar studies, not least because the technique was relatively easy and could be applied to virtually any species. It broke the monopoly held by the few genetically manipulable species that had dominated population genetics, such as *Drosophila*, corn, and laboratory mice. By 1984, 1111 species had been investigated using the technique, with an average of 23 loci and 200 individuals per species examined. As Lewontin notes, this approach is sometimes referred to derisively as the "find 'em and grind 'em" school of genetics, but it firmly established what would otherwise have remained a matter of speculation: that in most natural populations, about 30 percent of genes that code for enzymes and other soluble proteins are polymorphic, and any individual is heterozygous for about

10 percent of its gene loci. Lewontin and Hubby found that the figures for heterozygosity also hold for different geographic populations of the same species. They examined 18 gene loci in five populations of *Drosophila pseudoobscura*—Strawberry Canyon, Wilrose, Cimarron, Mather, and Flagstaff—from the southern and western United States. Their results, given in the accompanying table, demonstrate that the heterozygosity of a species remains constant from population to population and is not a matter of genetic differences between geographically isolated populations.

Even with the development of electrophoretic methods that could detect the previously invisible amino acid replacements, these figures did not change—about 10 percent of an individual's genes were heterozygous. Loci that were revealed as monomorphic remained so. Those that had been detected as polymorphic were found to have many more variants than formerly thought, sometimes dramatically so (an increase from 8 to 27 in the case of the enzyme xanthine dehydrogenase, for instance). When such studies were extended to other classes of protein, a similar picture emerged.

The development during the 1970s and 1980s of methods for reading the sequence of nucleotide bases that constitute the DNA molecule removed any lingering doubt about the extensive nature of genetic variability. This, the most fundamental level of an organism's genetic blueprint, is even more variable than is revealed in protein structure, for the simple rea-

Extent of Genetic Variation in Populations of
Drosophila pseudoobsura

Population	Number of proteins tested	Number of loci polymorphic	Percent of loci polymorphic	Percent of genome heterozygous per individual
Strawberry Canyon	18	6	33	14.8
Wildrose	18	5	28	10.6
Cimarron	18	5	28	9.9
Mather	18	6	33	14.3
Flagstaff	18	5	28	8.1
Averages			30	11.5

son that many aspects of DNA sequence do not affect the amino acid sequence of proteins.

Two levels of genetic variation, therefore, are of concern to population geneticists: the variation in gene products (that is, proteins), which are "visible to" natural selection because they affect the physical integrity of the organism; and the myriad variations in DNA sequence, much of which is "invisible" to natural selection because sequence changes often have no effect on protein structure. The question that has dominated much of population genetics—both theoretically and empirically—for the three decades since the existence of extensive genetic variation was demonstrated is, What is the origin of this variation?

Chance? Or Selection?

Once extensive protein polymorphism in populations had been demonstrated, there very quickly followed the development of two opposing explanations. The first assumed that it was the product of natural selection, and represented the active accumulation of variants that are adaptively advantageous under the many different environmental circumstances to which the population was exposed. In other words, a significant proportion of genetic mutations become fixed in the population, because they are beneficial. Deleterious mutations are removed, because the individuals possessing them are less fit. The second explanation viewed extensive genetic variation as simply the passive accumulation of chance mutations creating new alleles, the vast majority of which are adaptively neutral—they neither enhance nor diminish an organism's fitness. In this view, the principal role of natural selection is the removal of rare deleterious alleles.

To selectionists, therefore, most mutations are either beneficial or harmful; beneficial ones are retained in the population, creating extensive variation, while harmful ones are removed. To neutralists, most mutations are adaptively neutral, and therefore become fixed in the population because their presence poses no harm; extensive genetic variation is the result. This, simply stated, is the basis of the selectionist-neutralist debate.

Those favoring selection needed to find evidence that variation had beneficial effects, and they turned to two kinds of data, functional and static. Functional data document the existence of protein variants and describe how these correlate with different habitats. The idea was that if such a correlation exists, then it might reflect the fitness superiority of one variant over another under different environmental conditions. Static data are a measure of the overall level of variation within populations, and one issue was whether higher levels of variation equated with fitter populations. With the flood of electrophoretic data available to them, population geneticists fell to both tasks with gusto. The upshot of their effort was at best ambiguous, both in the static and functional realms. Some investigations yielded support for selection, but they were few—and certainly not the strong signal that was expected. At the same time, the neutralist view gained ground, not only because of the relative failure of the selectionist alternative but also because of its own elegance and power.

The seeds of the neutral theory (or, more properly, the neutral mutation random drift hypothesis) had already been sown at the time the Lewontin/Hubby/Harris data were being gathered. In 1965 Emile Zuckerkandl and Linus Pauling, then of the California Institute of Technology, published data on the evolution of hemoglobin molecules, obtained by comparing the amino acid sequences of hemoglobins from several species. Remarkably, these data indicated the following: amino acid substitution occurred at a constant rate; and the rate was high. When the Japanese geneticist Motoo Kimura, of the National Institute of Genetics in Mishima, saw these findings, they triggered in him a line of thinking that eventually led to the elaboration of the neutral theory, which he formally announced in November 1967 and published the following year. Together with similar data on the evolution of cytochrome c proteins from many species, the hemoglobin information appeared to offer a picture of the origin of variation that Kimura judged could not be completely explained by the process of natural selection. The level of variation seemed to be too high, and it accumulated too rapidly. When the first protein electrophoresis results showed that such variation was widespread, Kimura concluded that the most persuasive explanation was that the alternative alleles that accumulate in a population are largely selectively neutral—or, more properly, selectively *equivalent*. That is, they don't alter the functioning of the protein being coded for; they are thus invisible to natural selection because one is not favored over another.

Even when several equivalent alleles exist in a population, one of them may nevertheless become the more common, sometimes to the exclusion of the others, over a period of time. A selectively equivalent allele becomes *fixed* in the population in this way through chance processes, not selection. This is termed genetic drift.

Motoo Kimura developed the theory of neutral molecular evolution in the late 1960s. Initially unpopular, the theory has come to dominate ideas about the mode of molecular evolution.

The diagram represents a model of genetic drift, showing how random events can, with time, lead to a substantial change in the distribution of genes in a population.

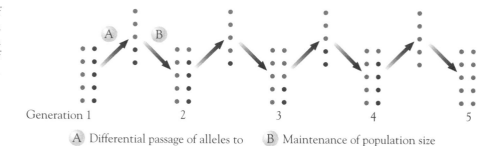

Generation 1   2   3   4   5

(A) Differential passage of alleles to next generation   (B) Maintenance of population size

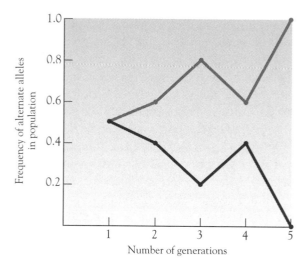

A thought experiment will illustrate the process. Imagine a bag that contains 100 balls, 50 blue and 50 red. Now imagine putting your hand into the bag to remove 50 balls, without selecting which ones you grab. You might randomly remove 25 blue and 25 red; but you are equally likely to remove, say, 30 blue and 20 red. Now imagine that the remaining balls replicate themselves, returning the total to a hundred. (This is the equivalent of a population of organisms remaining constant.) If 30 blue balls had been removed, the new population of balls would be 40 blue and 60 red. (This is the equivalent of "red" individuals producing more offspring than "blue" ones, by chance.) Repeat the process many times, and the relative proportion of blue and red balls will fluctuate because of these chance processes. Before long, however, there is a strong possibility that the population will constitute balls of just one color. The same thing happens in natural populations.

In 1969 J. L. King and Thomas Jukes, of the University of California, Berkeley, expanded the neutralist school with the publication of a paper in *Science*, provocatively titled "Non-Darwinian evolution: random fixation of selectively neutral mutations."

The neutral theory, it should be noted, does not preclude selection as a source of variation, but it does say its role is minor. As Kimura recently put it: "According to the theory, the great majority of evolutionary mutant substitutions at the molecular level are caused by random fixation, through sampling drift, of selectively neutral (i.e., selectively equivalent) mutants under continued mutation pressure. This is in sharp contrast to the traditional neo-Darwinian theory of evolution, which claims that the spreading of mutants within the species in the course of evolution can occur only with the help of positive natural selection."

An opposition of views was therefore quickly established. The real world that both were attempting to explain is one in which organisms seem to be well suited—that is, adapted—to the exigencies of their environments. In *Origin of Species*, Darwin had explained this adaptation by saying that "species have changed, and are still slowly changing, by the preservation and accumulation of successive slight favourable variations." Darwin's genius was to offer an explanation of how species come to be intricately "designed" to operate in their environments, through the naturalistic process of natural selection. At the end of the introduction to the *Origin*, he stated, "I am convinced that Natural Selection has been the main but not exclusive means of modification." He therefore allowed for other agents to contribute to the shaping of the world—including neutral evolution in today's formulation—but only as subsidiary influences. Darwin's view makes intuitive good sense, given species' adaptedness to their world: selection leading to adaptation *must* be important in evolution. And yet Kimura would turn this around. Paraphrasing Darwin, he wrote recently, "I am convinced that Random Drift acting on Neutral Mutants has been the main but not exclusive means of molecular evolution." The argument, then, is one of the *relative* importance of selection as against neutral drift.

Predictions of the Neutral Theory

Because of its mathematical simplicity, the neutral theory is able to make several firm, quantitative predictions about variation: namely, concerning the extent of variation expected in a population; the rate at which it accumulates, and the circumstances under which it might be expected to be maximal. A scientific theory stands or falls on how its predictions measure up to testing.

Under the neutral theory the *extent* of variation expected in a population is a simple function of the rate of mutation and what is termed the effective population size. When this equation is used to calculate variation, or heterogeneity, in most populations, the result is typically higher than the variation actually observed. Supporters of the neutral theory respond to this discrepancy in several ways. One response, developed by Tomoko Ohta, Kimura's longtime colleague, is to suggest that rather than viewing alleles as strictly neutral, we should judge them as nearly neutral, or slightly deleterious. Selection would remove these deleterious alleles, thus reducing heterogeneity in the population. Even under this modified formulation of the neutral theory, calculated variation *still* exceeds observed variation.

Another possible explanation for the discrepancy is that effective population size is consistently overestimated, artificially increasing the expected variation. If a population remains the same size through long periods of time, the effective population size is the same as the actual size. In reality, however, population size often fluctuates through time, sometimes dramatically so, as a result of disease or detrimental changes in habitat, for example. The effective population size is therefore some kind of average of these fluctuations. The upshot of this and other factors is that the effective population size is almost certainly lower than the actual size, often substantially so. If true, then this discrepancy might explain why the extent of genetic variation observed in populations is lower than expected. It is very difficult to prove this possibility, however.

The results of these tests of the theory's first prediction highlights an irony in the selectionist-neutralist debate. While selectionists were trying to understand how the extent of variation could be as *high* as is observed, neutralists were pondering why it is so *low*.

The issue of *rate* of accumulation of genetic variation is central to the neutral theory. As we will see later, it leads directly to the notion of a molecular evolutionary clock. As the neutral theory would have it, the rate of accumulation of variation is simply determined by the rate of mutation (that is, fixation of neutral alleles) for a particular molecule. (Recall that the notion that variation accumulates at a constant rate, driven solely by the rate of mutation, is antithetical to the Darwinian view of change, in which selection is the ultimate arbiter, keeping or rejecting the rarely neutral products of mutation.)

Molecules differ in the degree of modification they can tolerate: for instance, histones (proteins that help organize DNA molecules in the chromosome) have little tolerance for structural variation and, therefore, have a low rate of fixed mutation; hemoglobin, by contrast, can tolerate considerable change (at least in parts of the molecule) and therefore has a much higher rate of mutation. The differing tolerance for mutation has several implications. First, the neutral theory predicts that different genetic loci within the same organism will accumulate mutations at different rates. Second, it predicts that the same gene among different species will have an equivalent rate of fixed mutation.

Such strong predictions should be readily available for falsification, and in some observers' opinion they have been. This is one of the neutral theory's more vulnerable points, because strict metronomic clock behavior is not observed. The rate of change in the same protein in different species is often observed to be different. It is also true to say, however, that the degree of clocklike behavior that has been observed is far higher than would be predicted under the selection model.

The third prediction also relates to rate of change, but is more specific and, again, eminently testable. Early on, Kimura described molecular evolution under the neutral theory as being very conservative. By this he meant that functionally important molecules or parts of molecules will change less quickly than unimportant ones. This view presents a challenge for the Darwinian view: if selection is the driving force of evolution, then rates of evolution should be fastest where selection is operating most. From a Darwinian perspective, highly functional molecules or parts of molecules ought to change the *most*, since function is where selection exerts its pressure. This prediction is the opposite of that made by the neutral theory.

Natural selection has to protect what works, of course, weeding out mutations that interfere with a molecule's function. Its doing so will slow down the accumulation of changes. If mutations are neutral, however, leaving function unaffected, they are invisible to selection and can accumulate at a maximum rate. The question is, therefore, whether the observed maximum rate of change better fits the predicted effects of selection or the random accumulation of neutral alleles. The answer, unequivocally, has been the latter, and represents strong support for the neutral theory.

	24	25	26	27	28	29	30	31	32	33	34
L. PICTUS messenger RNA	GAU	AAC	AUC	CAA	GGA	AUA	ACU	AAA	CCG	GCA	AUC
L. PURPURATUS messenger RNA	GAC	AAC	AUC	CAA	GGU	AUC	ACG	?	?	GCU	AUC
Histone IV amino acid sequence in both species	Asp	Asn	Ile	Gln	Gly	Ile	Thr	Lys	Pro	Ala	Ile

The third nucleotide position in a codon can be mutated, usually without affecting the amino acid for which the codon codes. It is known as the silent site, and mutations there are called synonymous mutations. The chart illustrates the high rate of such mutations by comparing, in two species of sea urchin, a short stretch of messenger RNA coding for sites 24 through 34 of the protein histone IV.

What Drives the Rate of Change?

The first definitive evidence in support of this prediction of the neutral theory came from looking at rates of change in the three nucleotides that make up a codon. Each codon specifies an amino acid, so that a string of codons specifies a string of amino acids that form a protein. The important discovery here was what happened when a mutation caused a substitution of nucleotides: such a substitution had different consequences, leading mutations to accumulate at different rates, depending on the position in the codon. Specifically, nucleotide substitution in the first and second position in the codon almost always results in the encoding of a different amino acid. By contrast, changing the nucleotide in the third position typically does not change the amino acid specified by the codon. The third position is called the silent site, and a change to the third position is called a synonymous substitution. It is now known that mutations accumulate at such sites at a rate approximately twice that observed in nonsynonymous sites, just as the neutral theory predicts.

Some of the more intriguing discoveries of recent decades in molecular biology, beginning in the mid-1970s, have revealed segments of DNA that are *not* functionally constrained—that is, contribute no information to the final protein product—and therefore should be subject to high rates of substitution, according to the neutral theory. The first examples are known as introns (see the box on pages 42–44); these are segments of DNA that interrupt the coding (functional, information-containing) regions of a gene. The coding regions are called exons. During transcription of DNA into messenger RNA, the introns are edited out and (for the most part) disposed of. Mutations accumulate more quickly in these noncoding introns than in exons, and under some circumstances at a rate comparable to that observed in the silent third position of codons.

The second examples of functionally unconstrained segments of DNA are known as pseudogenes, or dead genes. Such genes are derived from functional genes by the process of gene duplication, as described earlier. Frequently, duplication produces copies of genes that lack introns and associated regulatory sequences. Such a duplicated gene cannot specify a protein, hence the term pseudogene. In the early 1980s, studies of a globin pseudogene revealed rates of substitution five times that observed in functional globin genes. Moreover, there was no difference in rates between the first two positions and the silent third position in pseudogene codons.

A particularly interesting example of what happens when a gene is released from functional constraints was reported in the late 1980s. This is the alphaA-crystallin gene in the Near Eastern mole rat. In many animals, the gene codes for a protein used to make the eye lens. Mole rats of this species, *Spalax ehrenbergi*, live a subterranean life, rarely, if ever, seeing the light of day. Their eyes develop to a rudimentary state, but the animals are completely blind. The neutral theory predicts that, since its gene product is not needed, the alphaA-crystallin gene should accumulate mutations at a higher rate than equivalent genes in sighted species of mole rats. This was demonstrated to be true: the rate is four times higher. However, it is not as high as the rate observed in pseudogenes, perhaps understandably. After all, the blind mole rat's alphaA-crystallin gene is expressed (that is, transcribed and translated into a protein), whereas pseudogenes are not. The alphaA-crystallin gene of the blind mole rat is part of a coordinated, developmental set of genes that produce a defined structure—the eye—however rudimentary. There is therefore *some* functional constraint on the gene, perhaps accounting for its reduced rate of mutation compared with that of pseudogenes.

A final piece of evidence in support of the neutral theory comes from RNA viruses; these viruses, which include the virus that causes influenza, employ RNA as their genome, rather than the DNA typical of all other organisms. Recall that the theory states that because most mutations are selectively neutral they will accumulate in the population, rather than being weeded out by selection. (Genetic drift will then sort the variants, by chance.) This means that mutations will initially accumulate in the population at a rate that very nearly matches the rate at which they occur. This

Near Eastern mole rats, which live in dark caves, develop only rudimentary eyes. The gene that codes for alpha A-crystalline, a protein in the eye lens, is released from constraints on mutation, but not completely so.

effect is known as mutation pressure. By contrast, selectionists hold that mutations accumulate only when they are selectively more favored than existing variants. Under the neutral theory, therefore, organisms that have an intrinsically higher rate of mutation will also have a higher rate of accumulation of new variants. RNA viruses have a much higher rate of intrinsic mutation than DNA viruses. The observation that RNA viruses accumulate mutations at a similarly higher rate is therefore consistent with the neutral theory.

In sum, according to Kimura, molecular evolution contrasts with anatomical evolution in two ways—its constant rate and conservative nature.

Selection and Neutrality Find Their Own Levels

When the neutral theory was first proposed, it drew heavy criticism from population geneticists because it was so at variance with the selectionist's view of evolution. Nevertheless, within a decade it had not only become firmly established as a serious alternative to the selectionist theory, but also emerged as the dominant theory. For instance, John Gillespie, an evolutionary biologist at the University of California at Davis and one of the theory's strongest critics, wrote the following assessment a decade ago: "The neutral theory . . . has had an enormous impact on population genetics, molecular biology, and our ideas about evolution." Effectively, the neutral theory became the null hypothesis of molecular evolution—that is, the simplest way to interpret genetic variation at the molecular level and the hypothesis whose predictions must be shown to be incorrect before the alternative (balancing selection) should be seriously considered.

Gillespie and others continue to criticize the theory, however, principally because the rate at which genetic variation accumulates has so often been observed to deviate from constancy. Gillespie argues that the observed rate is better explained as the result of occasional bursts of accumulation of substitutions driven by selection, not by steady accumulation. Kimura describes this scheme as "highly unrealistic," arguing that there is sufficient observed clocklike behavior in the molecular world to require postulating that selection would have to have occurred with the same probability in different lineages. The argument is unresolved, although the neutral theory does have a majority of support among geneticists.

Gillespie and other critics of the neutral theory express incredulity at the notion that a significant proportion—say 10 percent—of nucleotides can be changed in a genetic locus with no impact on function. "Such has been the message from the neutral allele theory," observes Gillespie. Yet if the theory is so terribly wrong, he asks, "why is it the most widely held theory of molecular evolution?" There are two reasons. First, natural selection is exceedingly difficult to define at the molecular level, particularly in light of the extreme complexity of the genome and its mechanisms of change. Second, in the realm of mathematical description, the neutral theory has a clear advantage. It can be described by simple, testable equations, whereas the selection theory so far eludes such description.

Perhaps we can gain perspective by returning to the earlier quotation from Darwin, and Kimura's paraphrase of it. Recall that Darwin proposed that natural selection was responsible for most, but not all, evolutionary change, while Kimura suggests that neutral evolution is responsible for most, but not all, such change. Let us look at the evolutionary context here. Darwin was talking about *organisms* in their worlds, and their fittedness to those worlds. Kimura was talking about organic *molecules* (particularly DNA) in their worlds, and how the molecules change. Each formulation deals with a different world.

Kimura is very probably correct to claim preeminence for mutation-driven, neutral evolution at the level of the gene. But at higher levels—organisms and the populations in which they live—selection evidently plays an important, perhaps preeminent, role in shaping evolution. The relative contributions of selection and neutrality are therefore different at different levels of evolution.

Population geneticists therefore know two important facts about variation at the level of the gene: it is extensive, and it is driven to a significant degree by chance. The next issue is the *tempo* of the accumulation of variation.

*T*he horseshoe crab, a so-called living fossil, has not changed
morphologically in tens of millions of years. The existence of
such species suggests that the tempo of morphological evolution
is highly variable, even though the underlying genes may
accumulate mutations at a more regular rate.

The Molecular Evolutionary Clock

*T*he molecular evolutionary clock is one of the simplest and most powerful concepts in the field of evolution. It is also among the most controversial. Briefly, the clock concept is this. As species diverge from a common ancestor, they accumulate mutations at a *regular* rate, progressively becoming more different from each other genetically. By comparing the genetic difference between two related species, therefore, one can in principle calculate the length of time that has elapsed since they shared a common ancestor. The clock therefore adds a temporal dimension to phylogenetic trees that show relatedness among species.

Chapter

5

After a speciation event, genes will accumulate mutations independently in the separate lineages. Here, mutations are seen in a gene A, which diverges in sequence over time, giving A and A′. Even though the rate of mutation is not constant and equal in the two lineages, an average rate emerges that is similar; here, it is 5.5 changes in the time period represented; the total divergence between A and A′ is 11 mutations.

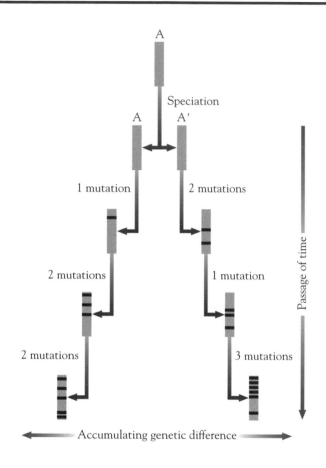

Biologists who would use the clock face a dilemma. They have demonstrated that it works (sometimes), but they do not fully know what makes it tick. It is as though you've successfully been using that venerable old time piece above City Hall for many years to order your daily routine. One day you decide to take a look at the mechanism that has been turning the clock so reliably (well, almost all the time) these many years. You climb the steps to the clock tower, throw open the doors to the clock's housing, and see nothing but pigeon feathers, dead rats, and other unspeakable debris—nothing that could possibly turn the hands of the clock in a regular manner. Naturally, you conclude that the clock cannot work after all.

Except, of course, you know that it does. This analogy stretches reality only a little.

The notion that the same protein molecules in related species are different from each other—and the corollary that the greater the evolutionary separation the greater the difference—has a long history. In the first decade of this century, E. T. Reichert and A. P. Brown, of the Carnegie Institute of Washington, compared the crystal structures of the protein globin from different species. Species of the same genus and sometimes the same family share structural characteristics, they found, but those in different genera and families do not. At about the same time, the British biologist H. F. Nuttall showed that immunological properties of certain blood proteins become progressively more unalike with increasing phylogenetic difference. He was interested in, among other things, the relationship of humans to monkeys and apes. This line of investigation was continued in various laboratories intermittently through the 1960s. For instance, in the early 1960s, as we have seen, Morris Goodman, of Wayne State University, used immunological properties of blood proteins to show that the African apes (chimpanzee and gorilla) are closely related to humans, while the Asian ape (orangutan) is equally distant from the three other species.

All these approaches assumed that homologous proteins (that is, proteins with the same evolutionary origin) in different species accumulated mutations as evolutionary time passed. But there was no assumption that the rate of accumulation was regular. This approach demonstrates conceptually that genetic differences among species can be used to reconstruct evolutionary histories, *whether or not* the rate of accumulation of mutations is regular, provided the fluctuations in rate are not wildly erratic.

The notion that the rate might be regular (that is, clocklike) was developed by Emile Zuckerkandl and Linus Pauling in the early 1960s, at the California Institute of Technology. In collaboration with Richard T. Jones, a student of Pauling's, Zuckerkandl treated the family of hemoglobins from several species with a digestive enzyme (trypsin) that broke the molecules into fragments; Zuckerkandl and Jones separated the fragments in two dimensions on a gel. Where the pattern of fragments differed between proteins or between species, they knew, so did the amino acid composition and sequence. In 1960 Zuckerkandl and Pauling reported that this technique indicated a close evolutionary relationship among humans and African apes, with the Asian ape more distant. This was several years before Goodman reported the same conclusion based on his work using immunological research techniques.

Left: Emile Zuckerkandl is a pioneer in the field of molecular evolution. Among other innovative lines of thinking in the field, he developed the concept of the molecular evolutionary clock. Right: Linus Pauling, winner of two Nobel prizes, worked with Zuckerkandl while both were at the California Institute of Technology.

This early work using tryptic digests, and more particularly data from studies to determine the amino acid sequences of hemoglobins (which had begun in the mid-1950s), led Zuckerkandl to conclude that the number of amino acid substitutions was directly proportional to the time span since evolutionary separation. In the hemoglobin chains, the rate was about one amino acid substitution per million years. Zuckerkandl based his age calibrations in this estimate on major evolutionary divergences known from the fossil record.

The notion of a molecular evolutionary clock was therefore formulated by early 1961, although not formally explicated as such until 1965, in a paper by Zuckerkandl and Pauling. This publication, remember, was influential in prompting Motoo Kimura's development of the neutral theory two years later.

The suggestion that the evolution of proteins proceeds in a regular, clocklike fashion was not well received, for good reason. At the time biologists believed that *nothing* in evolution displayed regularity. Their belief was based partly on observation and was partly a reaction to an early, discredited theory of the trajectory of evolutionary change, known as orthogenesis.

In the early decades of the century, some biologists had viewed evolutionary change as driven inexorably and regularly in a particular direction, sometimes with unfortunate results. Saber-toothed tigers, for instance, were said to have become extinct because their long canines continued to elongate generation by generation, eventually preventing them from closing their mouths. Certain oysters (*Grypea*) were said to have evolved themselves out of existence by coiling their shells once too often, slamming themselves shut in a prison of their own making. Orthogenesis did not always lead to dire results, however. Human "superiority" and "dominance" over the rest of nature were viewed as the inevitable outcome of an evolutionary trajectory to ever-greater intelligence.

By the 1950s the concept of orthogenesis had been discredited, not least because it was quite at variance with the observed world of nature. The tempo of morphological evolution in different lineages is self-evidently extremely variable, being fast in some, slow in others. The existence of so-called living fossils—organisms such as horseshoe crabs that have not changed morphologically for tens or even hundreds of millions of years—is sufficient evidence alone to falsify orthogenesis.

By the early 1960s, therefore, the tempo of morphological evolution was viewed as highly variable, both *among* and *within* lineages. And as Zuckerkandl recently observed, "The diversity of rates in morphological

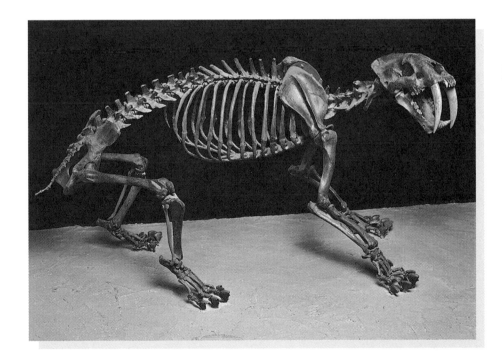

The saber-toothed tiger was erroneously thought to have evolved itself to extinction through the inexorable development of ever-larger canine teeth. The putative process that was thought to drive this inimical change was called orthogenesis.

evolution was the strongest argument, in the minds of biologists, against seriously considering the possibility that graded differences between informational macromolecules [proteins and DNA] might be proportional to evolutionary time." In other words, it was widely held that there can be no molecular evolutionary clock, because evolution does not work in a clock-like manner in any realm then observed. Not appreciated at the time was the distinction between structural genes, which code for proteins and various RNA molecules that operate in protein manufacture, and regulatory genes, which orchestrate overall gene activity. Mutations in regulatory genes are likely to lead to substantial morphological changes in the organism, as we saw earlier. Structural genes, on the other hand, may incur substitutions with little or no impact on morphology. The tempo of morphological evolution might therefore be highly erratic, as a result of occasional changes in mechanisms of regulation, while the genes (both structural and regulatory) that underlie morphology might accumulate mutations at a regular rate.

After the work on hemoglobin evolution, the next major application of molecular data to phylogeny made use of cytochrome c, a protein involved in energy metabolism in all forms of life. In 1967, Walter Fitch and Emmanuel Margoliash, then at the University of Wisconsin, compared the amino acid sequences of cytochrome c from 20 organisms, including humans, monkeys, ducks, a rattlesnake, a fish, and several microorganisms. They were able to reconstruct a phylogenetic tree that, with a few exceptions, matched what had been deduced from comparative morphology—a remarkable feat for scientists relying on information from just one protein. This work represented a landmark in the development of molecular phylogeny. The first real test of such data as a molecular clock, however, came in the same year when Allan Wilson and Vincent Sarich, of the University of California, Berkeley, extended Morris Goodman's work on human-ape relationships. Using a technique similar to Goodman's, Wilson and Sarich went beyond phylogeny, which purports to deduce only the *shape of the tree*. They calculated the *length of the branches* (representing time), by calibrating genetic distance against time span as reflected in the fossil record. Wilson and Sarich concluded that the human family had diverged from a last common ancestor with the African great apes about 5 million years ago.

As recounted in more detail in an earlier chapter, this proposal was highly unpopular. Anthropologists at the time considered the divergence to have taken place at least 15 and possibly 30 million years ago, according to

Allan Wilson was an innovative pioneer in the application of molecular techniques to biological questions.

their contemporary interpretation of the fossil record. The argument was also made that Wilson and Sarich could not be right, because there was no basis (then recognized) for believing in the validity of a molecular evolutionary clock.

During the subsequent decades, many researchers have addressed the same phylogenetic issue, using genetic information from a variety of sources, including proteins and DNA. Using molecular clock calculations, each produced a time for the human-ape divergence close to Wilson and Sarich's original claim. Meanwhile, anthropologists, having reevaluated existing fossil data and assessed newly discovered fossils, now consider the divergence date to be close to 5 million years ago. Here, apparently, *was* a molecular clock (in reality, a series of clocks: each protein and each DNA examined is its own clock).

If molecular evolution can proceed in a clocklike manner, as appears to be the case here, it is important to know what mechanism underlies such a progression. The neutral theory of evolution, as described in the previous chapter, provides such a mechanism and, with it, the predictive power that makes the mechanism scientifically testable, Kimura recently described the situation as follows: "From the standpoint of the neutral theory of evolution, it is expected that a universally valid and exact molecular evolutionary clock would exist if, for a given molecule, the mutation rate for neutral alleles *per year* were exactly equal among all organisms at all times." Deviations from this hypothetically equal rate of neutral mutation—caused either by a change in the rate of mutation (perhaps as an effect of differences in generation length, for example) or by changes in natural selection that, for instance, might make a sequence adaptively beneficial where previously it had been neutral—would make the clock less exact.

It should be noted here that even if mutation proceeded according to strict neutrality, the molecular evolutionary clock would not be a metronomic clock, ticking regularly year by year. Instead, it would be a stochastic clock, following the probability of mutation in a particular molecule year by year. Averaged over time, stochastic clocks of this kind nevertheless can be extremely accurate. A metronomic clock ticks regularly: in this evolutionary context it might do so, say, once every thousand years, so that after the passage of 5 million years, 5000 ticks would have occurred, spaced regularly. A stochastic clock, by contrast, is not regular, at least in the short term. In this same example, it might record a mere 500 ticks in the first million years, 1500 in the second, 1000 in the third, 300 in the fourth, and

Vincent Sarich teamed up with Allan Wilson in the mid-1960s, and, amid considerable controversy, soon had anthropologists questioning their assumptions about the time of origin of the human family.

The molecular evolutionary clock does not tick like a metronome but fluctuates in rate over time. Measures of evolutionary change are therefore averages of change over time. It is a stochastic clock, which can nevertheless be very useful in evolutionary studies.

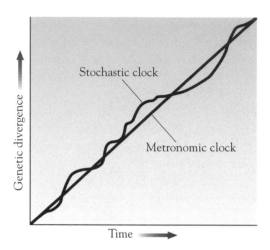

1700 in the fifth. By the end of the period, however, the stochastic clock has recorded the same number of ticks (or evolutionary changes in the form of mutations) as the metronomic clock, even though they accumulated irregularly. The average over 5 million years would thus be the same as the metronomic clock. An important point here, too often overlooked by proponents of the molecular clock, is that a stochastic clock becomes increasingly accurate as the time span being measured increases. The reason is the same as can be observed with flipping coins. After six flips, you are likely to have two heads and four tails, or 33 percent heads and 66 percent tails, for example; after a thousand flips, however, the distribution will be extremely close to 50:50, which is the statistical probability. With the passage of more time (more flips) chance biases are evened out. Logically, therefore, over short time spans, a stochastic clock might be quite inaccurate.

Even if strict neutrality in the accumulation of mutations does not apply—if, instead, selection plays an important, although fluctuating, role—clocklike behavior is still possible. Averaged over time, even a fluctuating mutation rate can yield a useful measure of time. An issue here is that of utility. If 99 percent temporal accuracy is required, a molecular evolutionary clock may not be up to the task; but if 80 percent accuracy is sufficient—as is often the case in biology where no other means exist for measuring time—then such a clock, however sloppy, may prove adequate.

It is certain, as we saw earlier, that different DNA sequences mutate at significantly different rates, often but not always because of different functional constraints—some, but not all, sequences may be altered without impairing the functioning of a protein. Sometimes, differences in the efficiency of DNA repair mechanisms lead to different rates of mutation. These differences in mutation rates can be seen at different levels in the genome. Within codons, for instance, mutations occur much more frequently at the silent third position than in the first two positions. Within a single gene, mutation rates will be lower in highly constrained functional regions and higher in regions under less constraint. In the globin genes, for example, mutations that affect the interior of the mature, globular protein are tolerated less easily than those that affect the protein's exterior, because the interior is the active site of oxygen binding.

Different genes within the same organism mutate at different rates, depending on how much of the coded protein can be modified and still perform its function: as noted earlier, histones tolerate little modification, hemoglobin much more, a difference reflected in their mutation rates. The genes for ribosomes, which are involved in the assembly of proteins, are also tightly constrained and mutate slowly. Introns and pseudogenes are subject to very little, if any, functional constraint, and they mutate at higher levels still.

Finally, different genomes in the same organisms mutate at different rates. While the nucleus is the major repository of genetic information in higher (eukaryotic) cells, the cytoplasm contains certain organelles that also possess genomes, such as mitochondria in animals and mitochondria and chloroplasts in plants. Chloroplast DNA evolves several times slower than nuclear DNA, whereas mitochondrial DNA typically evolves as much as 10 times faster than nuclear DNA. The differences here may well be

Slower   Faster

Ribosomal DNA Chloroplast DNA	Exon Intron Pseudogene (nuclear DNA)	Mitochondrial DNA

Different types of DNA accumulate mutations at very different rates, potentially giving clocks that are useful for evolutionary studies at different time depths. Ancient divergences are scrutinized best with slowly mutating DNA (such as ribosomal DNA), while recent divergences are best probed with fast changing DNA (such as mitochondrial DNA). Protein-coding genes in nuclear DNA accumulate mutations at an intermediate rate. In general, however, exons change more slowly than introns, and pseudogenes are freed to evolve quickly.

traced to different efficiencies of DNA replication and repair among the organelles and the nucleus.

These differences can, for the most part, be accommodated under the neutral theory. Moreover, the different rates at which mutations accumulate in different types of DNA may be exploited to investigate problems at different time scales. The investigator simply selects a molecular clock of the appropriate ticking rate. If, for instance, the phylogenetic question at hand requires comparisons to be made across hundreds of millions of years, then a very slowly ticking clock—ribosomal genes, for example—is most appropriate. Phylogenetic comparisons on a short time scale of, say, tens of thousands of years require a rapidly ticking clock, such as certain regions of the mitochondrial genome. And so on.

A Clock for All Species?

The central question with respect to the neutral theory itself (and as it pertains to the molecular evolutionary clock) is not whether different genes mutate at different rates but whether the *same* gene mutates at a constant rate among *different* lineages. Hundreds of proteins and genes have been sequenced in scores of species in pursuit of this question, which, in principle, would appear easily answered. Nevertheless, strong debate took place during many years over how to interpret the flood of data that was produced. Some argued that there was indeed a universal rate, both in nuclear and mitochondrial DNA, while others saw variability of rates. By now it is clear that rates *do* vary between lineages, although less so than some observers expected.

For instance, a survey of evidence from DNA-DNA hybridization comparisons among more than 20 species, undertaken in 1986 by Roy Britten, of the California Institute of Technology, found a fivefold difference in the rate of mutation. The rates of mutation are slowest in higher primates (particularly in apes and humans, or hominoids); the rates are also slow in some bird lineages. Rates are faster in rodents, sea urchins, and *Drosophila*. Britten saw a similar difference in rates when he examined data on silent substitutions in 25 genes among 29 species. Britten's survey of data may be taken as typical of the current consensus.

Why do rates vary? Britten's suggestion is that the groups with slow rates of mutation have more efficient DNA repair mechanisms, which weed out the mistakes that are made during the replication of DNA. Others have argued that species with very short generation times—such as mice compared with humans—will have higher rates of mutation, because there are more opportunities for mistakes to be made in the transmission of genes from generation to generation. It is true that the mutation rate in mice, say, is higher than in humans, but only fivefold so—well below the hundredfold difference in generation time. It has been suggested that the smaller difference in mutation rate relative to generation time is a consequence of the continual turnover of gametes (particularly male gametes, or sperm) that is always taking place whether species have long or short generation times. Sperm are being produced constantly in both species, allowing for mistakes to be made constantly, not just at the time of transmission.

Mutation rates are higher in species with high metabolic rates, suggesting that metabolic rate may also act as an influence on mutation rate. Recent support for this view comes from an analysis of certain mitochondrial DNA sequences in shark species. With colleagues from the University of Hawaii and the American Museum of Natural History, Stephen Palumbi calculated a rate of mutation seven- to eightfold *slower* than in primates and ungulates. Generation times in these various species is similar, but the metabolic rate is markedly different—five- to tenfold lower in sharks than in mammals of the same size. This study emphasizes the importance of measuring rates *within* groups rather than assuming it is safe to extrapolate from one group to others.

As part of their scrutiny of the molecular evolutionary clock hypothesis, biologists have surveyed which genes make good clocks and which do not. The task is to examine the gene (or its protein product) across many different groups of animals, looking for regularity in the accumulation of mutation—or its absence. Even here, things are not always what they might seem, for several complicating reasons. For instance, it has long been recognized that mutation rate in a given gene will inevitably appear to slow down as evolutionary time becomes substantial. The reason is that mutation, being a stochastic process, has a certain probability of occurring repeatedly at the same locus; this is called a multiple hit. As time passes, the number of multiple hits inevitably increases, reducing the apparent rate of mutation. Molecular clock calculations routinely take this complication into account, by statistical adjustment.

The reality is, however, more complicated still, not least because not all loci within a gene are equally vulnerable to mutation; and that

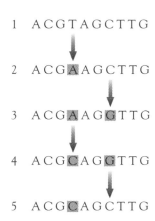

1 ACGTAGCTTG

2 ACGAAGCTTG

3 ACGAAGGTTG

4 ACGCAGGTTG

5 ACGCAGCTTG

The phenomenon of multiple hits potentially obscures the actual amount of mutation that has occurred. Here, four mutations occur in this short DNA sequence. At stage 3, an analysis of the sequence would accurately reflect the two mutations that have occurred by that time. The mutations that occur between stages 3 and 4 and between stages 4 and 5 will not be detected, however, because they hit previously mutated sites. Because the mutation between stages 4 and 5 reverts the nucleotide to the original one, it effectively obscures two mutations. An analysis of sequence 5 would therefore count just one mutation, not the four that have actually happened.

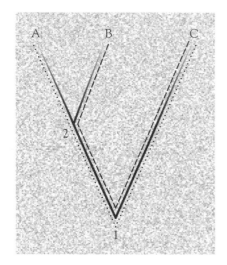

The relative rate test demonstrates whether of not the rate of accumulation of mutation in a protein or gene differs between different lineages. The diagram represents two evolutionary events. At 1, a split occurred, leading to species C on one hand and a second lineage on the other. This second lineage split at node 2, leading to species A and B. The rate test says that if the average rate of genetic divergence is the same in all lineages, then the genetic distance from species A to species C (dotted line) should be the same as the genetic distance from species B to species C (dashed line). If, however, gene mutation slowed down in lineage B, then the genetic distance from B to C would be less than that from A to C.

vulnerability itself may change, depending on mutations at other loci. Multiple hits will occur more frequently in a gene with some especially vulnerable loci than in one that is equally vulnerable throughout—again reducing the apparent rate of mutation.

Such complications may make what is actually a reasonably good molecular clock look rather bad. An example is provided by the superoxide dismutase gene, which produces a protein that (combined with copper and zinc cofactors) mops up toxic oxygen radicals in all aerobic organisms. A decade ago, Francisco Ayala compared the complete sequence of this gene in mold, yeast, fruit flies, a swordfish, a cow, a rat, and a human. He then applied these data to a phylogenetic tree of these organisms (which extends back 1.2 billion years) and calculated the mutation rate at different points in history and among different groups. The rates differed considerably. Among mammals, for instance, he found a rate of 27.8 amino acid substitutions per 100 residues per million years; between mammals and swordfish and between the swordfish-mammal group and *Drosophila* the rate was 9.1; and between fungi and animals, 5.5. Apparently the mutation rate of this gene varies fivefold, depending on the time span scrutinized. Ayala concluded that this gene "is a very erratic clock."

Ayala recently collaborated with Walter Fitch, now at the University of California at Irvine, in a reexamination of data for the superoxide dismutase gene, but this time they employed more extensive corrections for the complications of multiple hits. They found that this gene is in fact a rather *reliable* clock, and they recently wrote the following: "A reasonable model of the biological processes involved permits one to conclude that what at first appears to be a very inaccurate clock may be inaccurate simply because the necessary corrections have not been made." Similar conclusions may apply to other genes that have been described as erratic clocks. The need for appropriate statistical analysis is clearly demonstrated here.

By now it can be said that the molecular evolutionary clock is neither as good as had been hoped nor as bad as had been feared. Although its existence depends on there being an important degree of neutral evolution, even with neutral evolution many factors perturb what would otherwise be a universal rate and place any such clock very much in the empirical realm. Predictions about what may take place under untested circumstances are, therefore, very difficult to make. As empirical data build, however, predictability will become more certain. It is apparent that few "global clocks"—clocks that can operate over long time scales and a wide range of taxa—are likely to exist. Nevertheless, it is also clear that "local clocks" can be found and used reliably over limited ranges of time and lineages. (Their accuracy may be checked with the relative rate test, illustrated in the figure on the facing page.) It is also desirable to employ more than one gene, if possible, when making investigations using a molecular evolutionary clock. If scientists can demonstrate consistency among effectively different clocks, they will have greater confidence in any conclusions they may reach.

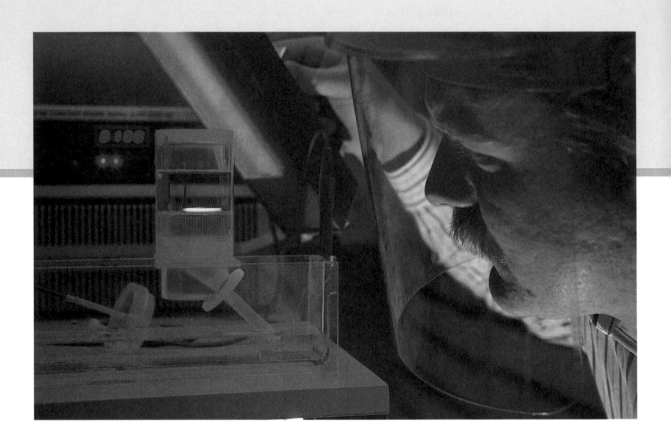

*T*hese days, an important endeavor of ecology takes place in
the molecular biology laboratory. Here, DNA is being extracted
from a tissue sample so that, for instance, the paternity of an
individual creature can be determined by DNA fingerprinting.

Molecular Ecology

*E*cology is an ancient and broad research discipline that explores issues from the interaction of a parasite and its host, for example, through the mechanisms of assembly of ecosystems, to the dynamics of global chemical cycles. Previously practiced under the guise of natural history, ecology has been building a picture of how the living world works, how organisms pursue their individual interests, and how species evolve. Molecular ecology, less than a decade old, was born of the application of new molecular methods to certain of these long-pursued interests.

Earlier chapters demonstrated how molecular studies have allowed us to resolve key evolutionary relationships: Which are the most fundamen-

Chapter

6

tal branches in the early tree of life? How are the major groups of multicellular organisms related to one another? What were the principal patterns of evolution within these groups? In the present chapter we will discover how these insights can be extended in two important directions. The tree of life is, after all, not only a genealogical record but also a vigorous and complex pattern of organisms that coexist at any particular time; those still flourishing represent the ecology of Earth today. Not only the tree of life, therefore, but also the web of life is subject to illumination by molecular genetic techniques. Beyond biological mapmaking—plotting taxonomic boundaries and historical divergence—we can now, in some instances, see how these branches arose, leading to fresh perspectives on why species diverge and new species form.

Chapter 4 described the unexpected discovery in the 1960s of tremendous genetic variation within species, as revealed by gel electrophoresis of proteins. This discovery not only prompted the neutralist-selectionist debate that continues today, but also permitted important studies of relationships among populations and species—forming the basis of population genetics. Since then, several key technical refinements have transformed scientists' capacity to distinguish not only crucial groups of organisms, but individual organisms themselves. One, for instance, was the development in the early 1980s of restriction fragment length polymorphism (RFLP) analysis, known more concisely as restriction enzyme mapping. As described in Chapter 1, the technique added power to the geneticist's tool kit, by giving access to genetic variation of populations at the level of DNA. Yet if genetic insights were to be extended more widely in ecology, a means of identifying individual organisms at the genetic level would be needed.

The relevant analogy here is the hundred-year-old technology of fingerprinting, which allowed human individuals to be identified with certainty. Such a means for identifying individuals of any species emerged in the mid-1980s, with the development of two techniques. First was DNA fingerprinting, which is similar in principle to the RFLP technique, but yields much more information about an individual's DNA.

The genomes of higher organisms contain many regions of DNA that apparently do not code for anything. It is sometimes called junk DNA, despite the fact that at least some such sequences might perform yet to be discovered functions. One such example is the repetitions of short sequences—the repeated, or core, sequence being some 15 nucleotides long—that are known as minisatellite DNA. There are many different

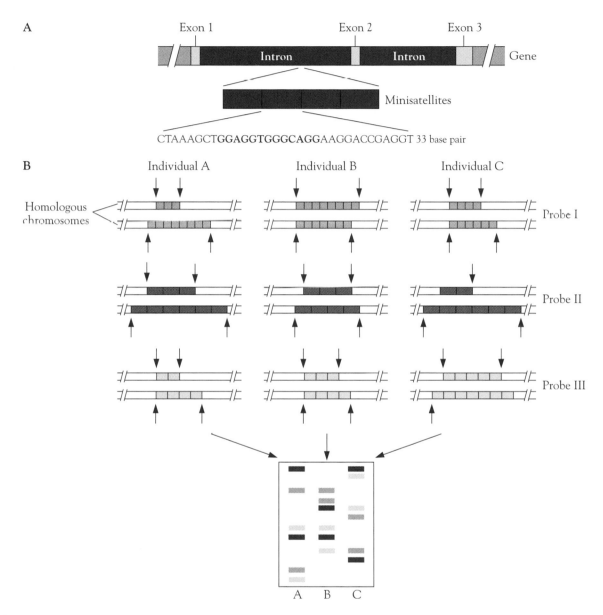

A

Exon 1   Exon 2   Exon 3

Intron   Intron   Gene   Minisatellites

CTAAAGCT**GGAGGTGGGCAGG**AAGGACCGAGGT 33 base pair

B

Individual A   Individual B   Individual C

Homologous chromosomes

Probe I

Probe II

Probe III

A   B   C

DNA fingerprinting identifies a unique pattern of DNA for each individual. The first step is the preparation of a probe (a). The first intron of the gene has four repeats of the sequence shown, in which is embedded the 13-base-pair sequence shown in boldface. This core sequence is found at other loci, labeled probe I, II, and III in this simple diagrammatic representation. The number of repeats at the three probe loci with the core sequence differs in the three individuals A, B, and C, providing DNA fingerprints (b).

The Power of PCR

In 1983 Kary Mullis, then at Cetus Corporation, helped develop a technique known as the polymerase chain reaction, or PCR, for which he received a share of the 1993 Nobel Prize in chemistry. The technique is as simple as it is powerful. It allows a researcher to produce large quantities of a desired piece of DNA, which can then be analyzed (sequenced, for instance) with ease. In principle, PCR can be applied to a starting sample containing just a single molecule of DNA and soon produce a billion copies. By 1990 the technique, which revolutionized molecular biologists' ability to obtain and analyze target segments of DNA in many types of investigations, had been cited in a thousand research reports; that figure doubled within a few years.

Prior to the development of PCR, obtaining a selected segment of DNA was a laborious business, due to the nature of the molecule and its physical context in the cell. A typical gene will have a combination of several thousand nucleotide bases, arranged like beads on a string in complementary sequences along the two strands of DNA that twist together to form the double helix. Inside the cell, the DNA double helix is further twisted into complex structures and is bound tightly to various proteins, which add stability. When a researcher tries to isolate DNA by breaking open the cell and disposing of the proteins, the long, delicate molecule frequently snaps at random places. This means that the same gene from, say, a thousand different cells will be found on sample strands of DNA of varying lengths, making isolation of useful quantities of the gene difficult.

In the 1970s, methods were developed to obtain repeatable, known lengths of DNA (using so-called restriction enzymes to cut the DNA strand), which improved the efficiency of isolation. Nevertheless, quantities were often small, and it was often necessary to boost the amount of material by inserting it into a bacterium on a replication vehicle (a naturally occurring piece of DNA, such as a plasmid). The bacteria were then cultured.

As each bacterium divides, making two copies of itself, the inserted DNA is copied, too. When sufficient material has been produced, the bacterial walls are dissolved, the target DNA is released from the vehicle by cutting enzymes, and the DNA is collected and purified. This time-consuming and technically demanding process, however, is not always successful or even possible.

PCR greatly simplifies the isolation and amplification of DNA samples. It is a cyclic process, with each cycle doubling the amount of target DNA sequence. Starting with one molecule, 30 cycles of PCR yields a billion.

The process begins with a crude separation of DNA from the experimental material, such as the root of a single hair. The PCR technique depends on knowing a short sequence (about 20 nucleotides) on either side of the target sequence; these short sequences, known as oligonucleotides, are then synthesized by standard techniques in the laboratory. The crude

DNA sample is now heated, which separates the double helix into two complementary single strands. The oligonucleotide primers (as they are known) are now added and the mix cooled a little; this allows the oligonucleotides to attach (hybridize) to the regions flanking the target DNA, because of their complementary sequences. In the presence of DNA polymerase, a natural enzyme that replicates DNA, and a mix of assorted nucleotides, the primer sequence is gradually extended across the target DNA, forming a short, now double-stranded piece of DNA.

This single cycle has therefore doubled the target DNA. By repeating the process of heating, adding primer, and replication, the extant target DNA is doubled again. Very soon, the length of DNA being replicated is simply the target DNA plus the flanking oligonucleotides on either end. Because of the power of geometric progression, a modest number of repetitions soon manufactures large quantities of the desired DNA. With PCR a researcher can achieve in a few hours what previously required days or weeks of labor.

When PCR was first developed, it required the repeated addition of DNA polymerase after each round of heating, because the heat that separated the DNA strands also inactivated the enzyme. Mullis realized that this step could be avoided if he used an enzyme that was able to tolerate the heat of strand separation. Many bacteria have adapted to high-temperature environments, such as hot springs, and they offered the possibility of the right kind of enzyme. He obtained one of these organisms, *Thermophilus aquaticus*, and enlisted Cetus workers to isolate its DNA polymerase. As expected, the *T. aquati-cus* enzyme tolerates the heating and cooling cycles, simplifying the technique. The whole process has now been automated, using a small, inexpensive machine.

The extreme sensitivity of PCR—its great strength—is also its greatest liability. The primers are faithful to the sequence to which they hybridize, but such sequences may coincidentally be present in "foreign" DNA, too. If such foreign DNA accidentally contaminates the reaction mix, unwanted sequences may be amplified. Scrupulous cleanliness is therefore necessary for faithful PCR work. Avoiding such subversion has been particularly crucial in research on ancient DNA, where the samples under study, such as dried skin, fossil bone, or insects in amber, can become contaminated with modern bacterial and even human tissue. On more than one occasion, a research team has triumphed at extracting putatively ancient DNA, only to discover that in fact the DNA derives from one of the researchers themselves, amplified from a few skin cells sloughed off onto the sample.

Given the simplicity of PCR and its immense impact in both basic and commercial molecular biology, it is curious that, at the time of its invention, all the components necessary for the implementation of the technique had been available for more than a decade. Any competent molecular biologist who had given it a few minutes' thought might have come up with the technique. Mullis himself, when he hit on the idea, was shocked that no one had developed it before. In its first decade, PCR—the product of those few inspired minutes—became a billion-dollar business.

Small quantities of DNA can be amplified to workable quantities by the polymerase chain reaction technique, which goes through repeated cycles of enzymatic DNA replication. Here, extracted DNA is being prepared for amplification.

minisatellites, each with its unique core sequence. Each of these genetic stutters is present in at least 20 locations of the genome, forming minisatellite families. Within each family, the lengths of individual members (that is, the number of times the core sequence is repeated at each location) may vary from individual to individual of the same species. Alex Jeffreys, of the University of Leicester, England, recognized here the potential for identifying individual organisms, and devised a method he called DNA fingerprinting.

As with the RFLP technique, the DNA is treated with enzymes that cut the strand at particular nucleotide sequences. The length of the fragments produced in association with each minisatellite location depends on the number of repeats of the core sequence at the location. Given that the minisatellites themselves vary considerably in length, and that each family has at least 20 such locations, the amount of information derivable from one minisatellite family is considerable. The technique yields a unique constellation of fragment lengths, and thus a unique genetic profile, for each individual.

The second development was the introduction in the mid-1980s of the polymerase chain reaction (PCR), which allows scientists to obtain workable amounts of DNA from tiny samples, such as a speck of blood, a hair root, a discarded feather, and even dried feces (see the box on pages 124–125). During the subsequent decade, many more fingerprint-like techniques were developed. Terry Burke, a zoologist at the University of Leicester, recently observed that "these and related techniques have broadened the scope of genetics in ecology well beyond the vision of the most far-sighted ecological geneticist." This newly available means of access to genetic data is the bedrock of molecular ecology, and the burgeoning volume of work it facilitated led to the establishment of a journal of the same name in 1992.

Molecular ecologists therefore now have many tools to hand, and can address many issues within the field's four main areas: evolutionary ecology, behavioral ecology, phylogeography, and conservation genetics. By way of illustrating the scope and power of molecular ecology, the remainder of the chapter will describe examples of studies in each of these areas. All are united in their technical approach, and most in their goal of understanding more deeply how the world of nature works, at many levels. Conservation genetics has the added dimension of seeking ways of saving species from extinction in the face of the human destruction of habitat.

Evolutionary Ecology

A central tenet of Darwinian theory is that individuals will behave in a way that maximizes their fitness, by which is principally meant their reproductive success. At the simplest level, individuals should behave so as to maximize the number of their own offspring. In recent decades, biologists have had to expand this view to account for so-called altruistic behavior, the efforts of an individual to support a close relative in its reproductive efforts, possibly to the detriment of its own. A young male baboon that helps a brother fend of aggressive attacks, thus endangering himself in the process, is one example here. He might also help his brother monopolize the attention of a female, thus boosting his brother's chances of mating with her. Because the individual shares half its genes with his brother, his brother's reproductive success is also his own, albeit the benefit he receives is more limited than the benefit from producing his own offspring. The individual's total reproductive success therefore reflects a combination of its own offspring and those of its close relatives—a total measure known as inclusive fitness. Field biologists have expended a great deal of effort in dis-

127

cerning mating systems in a wide range of species, in order to test the operation of this Darwinian imperative. Molecular approaches offer a means of extending these field observations significantly.

Males and females are engaged in a constant battle to produce as many surviving offspring as possible, while expending the least effort. It is in the interest of each parent to have the opposite-sex parent rear the offspring, so that he or she may have an opportunity to produce more offspring with other fleeting mates. This conflict has been resolved in many different ways, producing a wide variety of mating systems, which differ partly for reasons to do with the biology of the given species, and partly for reasons of ecology. For instance, because fertilization is internal in mammals, males frequently contribute little more than their sperm to the joint reproductive effort, leaving the females holding the babies; paternal care is rare and it is common for one male to monopolize several females (polygyny). Fertilization is also internal in birds, but because one parent by itself cannot provide all the effort needed to rear its young, paternal care is much more

In mammals, polygyny (one male monopolizing several females) is a common form of social structure, seen here in moose.

common than in mammals; it is also common for several males to attend one female (polyandry). In fish, fertilization is external, and in many cases the male is left to care for the brood.

Whatever the fundamental mating system, males and females can be expected to try to maximize their reproductive success by producing offspring outside the system; by sneaky mating (for males) and sneakily depositing offspring in the care of other parents (an option, open primarily to birds, known as egg dumping or intraspecific brood parasitism). This tactic of egg dumping is surprisingly common, and is practiced by the house wren, the eastern kingbird, and the cliff swallow, for example. In these species, up to half the broods contain eggs laid by other females. Females may also seek matings beyond their principal males, choosing individuals of high genetic quality, whom they often induce to join in provisioning the offspring.

Biologists have turned to molecular techniques to establish the frequency with which males and females are successful outside the basic mating system. Most of the initial studies focused on bird species, popular subjects in evolutionary ecology. Studies on other groups have since been carried out, with the collective discovery of three main points. First, mating systems are more varied than had been imagined. Second, so-called alternative mating (sneaky fertilization, brood parasitism, and so on) is more common than formerly suspected. And third, females exert more choice of mates (particularly outside the system) than had been thought. The case of dunnocks, already described (pages 1–2), revealed the degree to which males could match their rearing efforts to their reproductive success. Another early study, on red-winged blackbirds, produced many surprising results, some of which revealed the limitations of traditional field observations in estimating reproductive success.

The mating system in red-winged blackbirds (a North American species) is polygynous: a male defends territories occupied by several females on their nests. There is no paternal care. Traditionally, male reproductive success was measured by the number of nestlings produced in a given male's territory, even though the females were known to be approached by males other than the territory holder. (The resident male spends a lot of time monitoring his females so as to chase away such unwanted intruders.) Similarly, females were considered to make the choice of where to nest and mate, based on some unknown (to humans) quality of the resident male. A major study by Lisle Gibbs and several colleagues, at Queen's University, Kingston, Ontario, and Carelton University, Ottawa, revealed just how erroneous these conclusions were. The researchers were

Male red-winged blackbirds defend a territory occupied by several females. The males expend a lot of time keeping other males from intruding on the territory seeking sneaky matings. Males who are successful at defending their own females are also successful at achieving sneaky matings in neighboring territories. Here a male is shown at the left, photographed in Ogden, Utah, and a female at the right.

able to determine the genetic identities of many birds (111 nestlings, 21 putative male parents, and 31 putative female parents in 37 nests) by using several minisatellite families in their DNA fingerprinting work.

The genetic data showed that in 45 percent of nests, the genetic profile of at least one chick did not match with one of its putative parents, and 28 percent (31 of 111) of the chicks had genotypes not compatible with one putative parent. In all cases, the excluded parent was the male. Female red-winged blackbirds, therefore, unlike many bird species, do not indulge in egg dumping.

This degree of sneaky mating on the males' part was unexpected, as was its pattern. In almost all cases, the father of such chicks was the resident of a neighboring territory, which is perhaps not surprising. What was surprising, however, was the fact that males who were successful in their own territories (that is, who were efficient at *excluding* extra-pair fertilizations) were also successful in *achieving* such fertilizations elsewhere. Furthermore, there was no correlation between the size of a male's territory and his reproductive success there.

In this case, therefore, molecular genetic analysis not only overturned traditional views about males' reproductive success, but also questioned assumptions about female choice. That is, it suggested that a female's choice of where to nest may not be influenced by which male she will mate with. The analysis also raised the interesting though still unanswered question,

What was the quality possessed by those males that were successful both in their own territories and in those of their neighbors?

A recently reported study of superb fairy-wrens, an Australian species, is a good example of how molecular analysis can reveal unexpected complexity in a mating system; in this case, it brought to light the highest known level of extra-pair fertilization. Superb fairy-wrens live in cooperative social groups composed of a single female and one to four adult males that defend a permanent territory. When several males are present, one is usually oldest and behaviorally dominant; the rest are sons of the resident female that are unable to disperse because of a shortage of available females and habitat. All the males contribute to defending and provisioning the brood. The pairing between the female and the primary male is stable from year to year.

Andrew Cockburn and his colleagues at the Australian National University, Canberra, used DNA fingerprinting on all individuals in 65 broods over two breeding seasons. They found that most offspring, some 76 percent, were the progeny of males *outside* the social group. In almost half the broods, *all* the offspring were fathered by males outside the group. When a group consisted of only a single pair, the male sired up to 90 percent of the offspring, and he engaged in extensive parental care. As the size of the social group increased, however, the percentage of offspring sired by the primary male dropped dramatically, sometimes to zero, and the father(s) were almost always from a distant territory. Only rarely did helpers in a group

Superb fairy wrens, an Australian species, live in cooperative social groups composed of a single female and one to four adult males that defend a permanent territory. A male is shown at the left and a female with two nestlings at the right.

sire offspring in that group, even though they might sire offspring in other groups. Among the most surprising results was the discovery that some of the extra-group fathers were significantly more successful than others, and that the sons of such fathers were also successful. Whatever quality these males possessed was obviously heritable. Moreover, all matings were solicited by the females, not the males.

Female choice, clearly strong in this species, is responsible for the unusual mating system. Cockburn and his colleagues argue that when a female superb fairy-wren is with a single male she allows him extensive, but not exclusive, mating opportunities in return for paternal care. When helpers are present, she no longer depends on the care-giving efforts of the primary male and seeks more extra-group matings, particularly with successful males; she benefits because not only do the secondary males provide parental care, but also because she is likely to have sons that are successful at extra-group matings. This study is the first to demonstrate in birds that females that choose extra-pair mates also produce offspring that gain extra-pair fertilizations. The theoretical issues raised by these results are legion and include questions about the benefits a female gains from extra-group pairing, why primary males care for young to which they are probably unrelated, and what determines the great variation in extra-pair fertilizations among species.

Unexpected female choice was also recently identified in gray seals, which live at remote sites around the British Isles. As mentioned earlier, the majority of male mammals contribute little more than their sperm to the raising of offspring, and polygyny is common. The gray seal is no exception. During the fall, females come ashore to give birth to single pups, suckle, and mate. Males, too, come ashore at this time and compete with each other for territories within the colony, in which they apparently monopolize a number of females.

Bill Amos, of Cambridge University, and several colleagues recently used DNA fingerprinting to determine paternal relationships among offspring of 85 males and 88 females over several seasons. They found that 30 percent of each female's pups were full siblings, which implies a surprisingly high degree of mate fidelity. Two main explanations were possible. First, a female might mate with the nearest dominant male, and males and females might frequently return to the same colony site. Second, seals might frequently select previous partners. The genetic data ruled out the first possibility, leaving the second as the most likely explanation, surprising as it was.

Gray seals live at remote sites around the British Isles. During the fall, females come ashore to give birth to single pups, suckle, and mate. Males come ashore at this time and compete with each other for territories within the colony.

The identity of the fathers, too, was a surprise. The most reasonable conclusion based on field observations had been that the dominant males would have sired more offspring than they actually did. The females were apparently making a choice of mates, possibly because fidelity to a male makes him less aggressive and thus less likely to kill the pups. Once again, access to individuals' DNA revealed a pattern of behavior that was not apparent in—indeed, proved contrary to—field observations.

Amos and two collaborators from the University of Munich exploited this same access to determine the social structure of pilot whales. Biologists are now adept at unraveling social structure in large mammals, such as baboons and chimpanzees, by long-term observation of groups in the wild. We have already noted that most mammalian species have a polygynous mating system. In virtually all cases, when males reach maturity they leave their natal group and join a neighboring or distant group (chimpanzees are an exception, where females transfer). This behavior is thought to be a mechanism to avoid inbreeding. Whether pilot whales behaved similarly was unknown until recently, for the possibility of long-term observation does not exist for pilot whales, which for most of their life are out of reach of human study.

The pilot whale swims in large, cohesive groups (or pods) often containing more than 100 individuals, young and mature together, with almost equal numbers of males and females. This behavior has been exploited for centuries by fishermen who are able to herd the pod, a practice that continues now only in the Faroe Islands, Scotland. Amos and his collaborators exploited this exploitation, by analyzing DNA from members of these herds. They extracted DNA from all individuals in many complete pods, and analyzed highly variable minisatellite sequences, obtaining information on maternity, paternity, and the relatedness of individual to individual. A highly unusual social structure and mating system was revealed.

Neither males nor females left their pods at maturity, so each pod was essentially an extended family. Nevertheless, the males did not mate within their own pod, but with females from other pods. All males appeared equally successful at siring offspring. Amos and his collaborators interpret this unusual system as follows. If males' mating opportunities in other pods are not limited, their optimal strategy need not include paternal care of

Pilot whales live in large, cohesive groups (or pods) often containing more than 100 individuals, young and mature together, with almost equal numbers of males and females.

their own offspring. By remaining in their own pod, males could support their sisters' and half-sisters' efforts to rear *their* young, thus optimizing the males' inclusive fitness. Perhaps the males defend the pod and help in feeding, although neither behavior has been directly observed. In any case, the close genetic relationship of pod members explains the great cohesiveness of the group.

Fitting into the Niche

The idea that species adapt to their environments by natural selection is at the core of Darwinian theory, and, like Darwin, modern biologists look to the comparative method to investigate it. If similar species in similar environments share similar adaptations, then the link between environment and species' morphology and behavior is apparently confirmed. However, such a putative link may be confounded. If the species share a recent ancestry, their similarities may be a legacy of history, not adaptation. In other words, the species might have inherited their similarities from a shared ancestor rather than evolved them as specific adaptations to a particular environment. In order to determine the true nature of such links among a group of species, it is necessary to establish their evolutionary history, or phylogeny. As we've seen, this is not always easy on the basis of morphological characters. Molecular characters are sometimes easier to deal with.

Adam Richman, of the University of Oregon, and Trevor Price, of the University of California at San Diego, undertook such a task with eight species of insectivorous leaf warblers in the Himalayas of Kashmir, India. These small, green birds differ in habitat choice, size of prey, and feeding method. There are strong correlations between a species' morphology and its habits. For instance, larger species feed on larger prey species; species with relatively short toes breed in coniferous, as opposed to deciduous, woodlands; and species with wide beaks include a significantly higher proportion of flies in their diet. Are these correlations the result of a shared evolutionary history, or of independent adaptation by natural selection?

Richman and Price constructed a phylogeny of the warblers, by comparing a 910 base-pair sequence from the mitochondrial DNA cytochrome b gene in all eight species, plus other species for outgroup comparison. Ac-

Eight species of insectivorous leaf warblers such as this one, *Phylloscopus trochilaides*, live in the Himalayas of Kashmir, India. They display considerable adaptation to their different habitats and prey.

cording to their research, the answers to the above question are yes and yes. For instance, the three smallest species are closely related, as are the three largest, and the two intermediate-size species. A shared history is therefore probably playing a role in these broad, body-size categories. Nevertheless, once the effect of history is removed statistically, there remains a correlation between body size and prey size. In the large-body-size group, a species feeding on large prey is bigger than a species feeding on small prey, and so on. Moreover, there is little effect of history on correlations of habitat selection, feeding behavior, and morphology. These correlations appear to be true adaptive effects.

The phylogenetic tree shows that the group originated close to 3.5 million years ago, then rapidly diversified, increasing from one to eight species over a period of about a million years. The initial split was accompanied by a large change in body size; subsequently, body size changed little within the resulting groups. The next shift brought large changes in feeding morphology and related behavior, including prey size. Most recently, species within the group adapted their behavior and morphology to fit new habitats. This pattern of evolution, which may be quite common in other groups, also reveals the tempo of evolution: long periods of little ecological and morphological change, and occasional periods of rapid change, usually in one aspect of adaptation at a time.

Interdependent Lives

Enveloping the roots of most plants like a gossamer cloud are millions of weblike filaments (hyphae) of fungi. The plants and fungi have a mutually rewarding arrangement: each partner derives essential nutrients from the other. Discovered only recently, this symbiotic relationship is the keystone of a viable biosphere, although it is but one example of the importance of symbiosis in the biological world.

Of perhaps even more fascination, if not quite the same magnitude of ecological importance, is the intimate relationship between attine ants, an insect tribe of some 200 species, and the fungi upon which they feed. Each depends upon the other for survival.

A leaf-cutter ant carries a bit of the fungus cultivated by its species in underground fungus farms. There are some 200 species of these ants, which are the dominant herbivore of the Neotropics, harvesting about 20 percent of the fresh leaf biomass there.

137

These leaf-cutting ants are the dominant herbivore of the Neotropics, harvesting about 20 percent of the fresh leaf biomass there. The ants don't eat the leaves, however—at least not directly. Instead they take them back to complex underground nests, where they process them as food for large, carefully tended fungal gardens. The fungi thrive on this bountiful food supply, much more so than if they had to depend on the casual arrival of leaves on the wind; and the ants thrive on the nutrient-rich fungal tips. Several mysteries attend this symbiotic arrangement. First, how and when did it begin, and how did it then proceed through time? Second, what is the identity of the fungi involved? (In establishing its partnership with ants, the fungi stopped producing fruiting bodies, which biologists have traditionally used for taxonomic identification of fungi.) It was also initially unclear whether the fungi belonged to a single family, having evolved in concert with the ants over many millions of years.

Although the phylogenetic tree of attine ants has been well worked out using morphological characters, no such information has been available for the fungi on which they depend. Researchers from half a dozen universities in the United States rectified this omission recently by constructing an evolutionary tree for the fungi, using sequences of nuclear ribosomal DNA. They were able to identify the fungi involved by comparing its ribosomal DNA with ribosomal DNA sequences of free-living fungi. Most turned out to belong to the family Lepiotaceae, which includes among its species the parasol mushroom, while the rest belong to a distantly related lineage. These fungi are unique in their ability to break down fresh leaves into usable nutrients.

The evolutionary pattern was as follows. It is apparent that the initiation or "invention" of fungus-growing behavior was a rare event in the history of ants, having occurred only once, some 50 million years ago. The ants switched from the original fungal lineage only once as well. The fungi's special leaf-digesting abilities explain the ants' fidelity to these lineages. Although some of the more primitive attines have switched within the Lepiotaceae from time to time, the molecular analysis indicates that the most evolutionarily specialized attines have propagated the same fungal lineage for at least 23 million years. Each time a queen ant leaves the nest she takes with her a small sample of the fungus, with which she establishes a new garden in a new nest. This means that the millions of tons of fungi in millions of underground homes of these ants throughout South and Central America all derive from a single ancestral spore that existed 23 million years back.

Mysterious Migrations

Some behavioral ecologists seek to understand patterns in the lives of species that are *un*related to an individual organism's reproductive success. Once again, field observations have proved successful in identifying some interesting behaviors, such as chimpanzees' habit of regularly eating leaves that have little or no nutritional value (they turn out to have medicinal properties). Certain behaviors defy direct observation, however, and an understanding of them requires other sources of information, such as genetic data. Two such examples focus on aquatic species, the green turtle and the humpback whale.

Lumbering on land but graceful in the sea, green turtles are magnificent animals, an important aspect of whose life was, until recently, a mystery (in effect, two mysteries). Populations of this turtle live at numerous localities around the equator. Each population spends most of the year grazing in shallow feeding grounds until it comes time to breed, when a migration to breeding grounds begins. Once there, females repeatedly visit sandy beaches every two weeks over a period of two months, excavating deep

A female green turtle coming ashore to lay eggs in Taim, Brazil. These creatures migrate between feeding grounds and breeding grounds, sometimes covering thousands of miles in the process.

nests and depositing more than a hundred eggs each time. When egg laying is finished, a reverse migration ensues, and the cycle begins again. For some populations, the round-trip migration amounts to several hundred miles; for others, it exceeds several thousand. One population, for instance, spends March to December feeding off the coast of Brazil. The females migrate to Ascension Island in the mid-Atlantic for breeding, and then return to the South American coast, a round trip of 2400 miles.

Biologists have studied the Ascension Island population and others around the Caribbean intensively for more than three decades, tagging females for identification. Year after year, for the several decades of their breeding life, the females return with great fidelity to the same breeding ground, even when they share feeding grounds with other populations. For instance, of 28,000 nesting females tagged over the past 30 years at a large rookery in Tortuguero, Costa Rica, none has been seen at another nesting site.

Why are females so faithful in their choice of breeding site? Is it, as was suggested more than 30 years ago, because they are driven by deep instinct to return to the beach of their birth (the so-called natal homing hypothesis)? Or is the initial choice determined more by chance, with young females following the example of older individuals, thus establishing a lifelong pattern? (This latter is known as the social facilitation hypothesis.) Unfortunately, it has not yet proved possible to tag young turtles in a way that could survive the 30 years until they reach sexual maturity, which would allow the question to be answered.

Tied up with this question is another one: Why should the Ascension Island population undertake such a challenging journey each year? Twenty years ago the late Archie Carr, father of marine turtle research, proposed an answer. He pointed out that 60 to 80 million years ago South America and Africa were separated only by a narrow channel. Ancestors of the present-day Ascension turtles may have colonized the proto-Ascension Island that lay between the neighboring continents. As time passed, the continents drifted apart by a matter of inches a year. Continental drift drew the turtles' breeding ground ever more distant from their feeding ground. Driven by instinct, the turtles gradually extended their migration generation by generation, until its present grand odyssey. Unfortunately, there was no way to test the hypothesis.

During the past several years, Brian Bowen, of the University of Florida, and John Avise, of the University of Georgia, have been addressing these questions using molecular techniques—specifically, they used restriction site analysis of mitochondrial DNA. Mitochondrial DNA is especially appropriate for the investigation: always inherited from the mother, it reveals the history and behavior of female populations. Male migratory behavior, if different from that of females, will not confound the results. If the natal homing hypothesis were correct, there would be no flow of mitochondrial genes between different rookeries; each rookery would therefore be genetically distinct. By contrast, social facilitation would cause a mixing of mitochondrial lineages from different rookeries, and rookeries would therefore *not* be genetically distinct. The mitochondrial data are clear. Even though a rookery at Henderson Island, Florida, and one at Tortuguero are genetically very similar, all others in the Caribbean—and the Ascension Island rookery—are genetically distinct. These results give strong support to the natal homing hypothesis.

What of Carr's hypothesis about the Ascension Island population? If it had indeed been separate from all other turtle populations for 80 million years, the genetic difference between the Ascension turtles and others would have been extensive. In fact, although the Ascension Island population is genetically distinct, it separated from other populations considerably less than a million years ago. That is the date of separation implied by the extent of sequence difference, assuming a sequence divergence rate of 2 percent per million years. Bowen and Avise believe that the divergence rate may be considerably slower in this species, perhaps just 0.2 percent. Even at this slower rate, however, Carr's hypothesis is not upheld.

Even long-separated populations could display a relatively small genetic difference, of course, if occasional interbreeding had produced a gene flow between the populations. Despite the strong natal homing behavior, females do stray into neighboring rookeries, although not frequently. More likely, Bowen and Avise argue, rookeries are rather ephemeral on a geological time scale, persisting for centuries or a few millennia rather than millions of years. Even before human activity obliterated many known rookeries, natural climatic shifts would have made some sites unsuitable for nesting. For instance, when the Pleistocene Ice Age ended 10,000 years ago, sea levels rose as much as 300 feet. Beaches that might have persisted on the cone-shaped Ascension Island for almost 100,000 years would have rapidly disappeared, and new ones formed as the new sea level became established. As suitable rookery sites disappear and others come into being through geological time, turtle populations are forced to move. During these times of "musical rookeries" opportunities would arise for gene flow among populations.

If rookery sites are ephemeral, furthermore, a genetically fixed instinct for natal homing is unlikely. Instead, suggest Bowen and Avise, all green turtles have a capacity for learning cues to the location of their birth. This, not a hard-wired map in the brain, is what drives their odysseys.

What of males in this unfolding mystery? They spend most of their life at sea and are rarely seen. Mitochondrial DNA tells nothing of their history and behavior. To answer the question, Bowen and Avise collaborated with Stephen Karl, of the University of South Florida, who compared nuclear DNA sequences from individuals in several colonies. Again the colonies were seen to be distinct, but not as clearly as revealed by mitochondrial DNA data. This implies that males occasionally mate with females from other populations, and are therefore principally responsible for gene flow between populations.

Humpback whales also engage in seasonal migrations, from summer feeding grounds in temperate or near-polar waters to winter breeding grounds in shallow tropical waters, often covering more than 7000 miles in the process. Whale watchers can identify individuals, often by unique patterns of the tail (or fluke), and thereby gain information about the species' population structure. In the case of the humpback whale, observers have concluded that in the Pacific and Atlantic Oceans there are subpopulations

Humpback whales undertake seasonal migrations, from summer feeding grounds in temperate or near-polar waters to winter breeding grounds in shallow tropical waters, often covering more than 7000 miles.

that have their own feeding grounds but often share breeding grounds. Genetic analysis offered a way of understanding these migratory patterns.

Scott Baker, of Victoria University in Wellington, New Zealand, and several colleagues obtained skin samples (by dart biopsy) from 90 individuals in two feeding grounds (southeastern Alaska and central California) and one breeding ground (Hawaii) in the North Pacific and one feeding ground (Gulf of Maine) in the North Atlantic. Baker carried out a restriction site analysis and also sequenced a fast-evolving region of mitochondrial DNA from these individuals. He found a clear genetic difference between the populations in the feeding grounds, not only between the oceans (not surprisingly) but also within the North Pacific. The Hawaiian breeding ground population was a mixture of Alaskan and Californian individuals. The absence of any geographic barrier between the Pacific subpopulations led Baker and his colleagues to conclude that the population structure in the ocean is the result of "strong maternal traditions in migratory destinations." Social facilitation is apparently important here.

Comparing the genetic difference between the Pacific and Atlantic populations leads to further insight into the species' behavior. The Isthmus of Panama separated the two oceans about 3 million years ago. The genetic difference between the Pacific and Atlantic populations should reflect that separation, if they have remained separate since then. Based on a rate of sequence divergence of 2 percent per million years, the difference between the populations should therefore be 6 percent. In fact, it is only a fraction of that—some 0.27 percent.

Either the rate of sequence divergence in humpback whales is very much lower than observed in other large mammals, *or* gene flow occurs between the oceans from time to time, possibly when subpopulations mingle. As with green turtle ecology, the feeding and breeding grounds of the humpback whale are also likely to be ephemeral, creating opportunities for such mingling. This work shows how, as Stephen O'Brien, one of Baker's collaborators, recently put it, "the phylogenetic analysis of a maternally inherited genome led directly to insight into the behavioral migratory strategies used by a large endangered marine mammal."

Phylogeography

Genetic variation is often not uniformly distributed across a species' entire range. Instead, discrete populations may be genetically similar within themselves, but distinct to varying degrees from other populations. In the case of green turtles and humpback whales, for instance, female migration keeps populations separate, thus allowing genetic differences to accumulate within the different populations. In these animals, such distinctiveness arose from the behavior of a species (migration), but it may arise for other reasons, too—particularly through a population's adaptation to local ecological circumstances. In this case, the genetic differences within a given species are viewed as a fine tuning of gene activity by natural selection. Isolated populations of the same species may also develop genetic distinctiveness through chance, or genetic drift, as we saw in Chapter 4.

There is a further agent for generating genetic distinctiveness among the populations of a species: history. History in this context is a complex process—an interplay of geography and climate. Geography determines whether or not a species can disperse to form separate populations, and climate determines whether a particular species can inhabit a given geographical region. Some organisms, such as tortoises, are very restricted in the territory they can cover from day to day and in their lifetime. As a result, there may be no breeding between individuals in neighboring populations. Gene flow is therefore limited between these populations, allowing genetic distinctiveness to develop over time. Mobile species—including some birds, marine fishes, and large terrestrial mammals—may form a large, effectively continuous breeding network, thus promoting gene flow among populations and limiting the establishment of genetic distinctiveness. Physical barriers to dispersal have the same effect as a species habit of limited dispersal. Ancient mountain ranges, for example, or major rivers allow substantial genetic distinctiveness to accumulate between populations on either side. Ultimately, such distinctiveness may become so great that the separated populations diverge to new species. Physical features of the landscape therefore imprint themselves on the genetic structure of populations living there. Genetic differences exist in geographically distinct human populations, for instance, although such differences are not great compared with the degree of genetic variation that exists within populations.

In 1987, John Avise and his colleagues coined the term phylogeography to denote the study of the impact of species' physical environments on their genetic structure. Avise argues that phylogeography should take its place with "ecogeography" to form collectively what ecologists know as biogeography. Ecogeography, for example, offers several "rules" to describe how populations of a species will manifest variations in anatomy in different ecological zones. According to Bergman's rule, for example, the bodies of warm-blooded organisms become bulkier (have a higher ratio of volume to surface area) at higher latitudes. Similarly, the limbs of such organisms become relatively shorter at higher latitudes, a correlation known as Allen's rule. Eskimos, for example, have bulky bodies with relatively short limbs, while human equatorial populations typically have tall, slender trunks and long limbs. Both adjustments are local adaptations to the demands of thermoregulation.

The introduction of mitochondrial DNA data to population genetics in the late 1970s provided a powerful tool for examining questions of phylogeography. Mitochondrial DNA is particularly suited to the problem because it is sensitive to genetic differences at the population, not the individual, level. As a result of such work, phylogeography has become recognized as a significant factor in the equation shaping genetic structure both within and between species. Analysis of such substructuring therefore gives information about a species' behavior (in terms of dispersal) and its history (how a species is distributed over the landscape, and has been for some considerable time). Some examples will illustrate these principles.

Avise and his colleagues were among the first workers to gather substantial phylogeographic data for a wide range of species. Their analysis of the pocket gopher, a burrowing rodent that lives across a three-state region in the southeastern United States, remains classic. Published in 1979, it relied on data from a restriction fragment length polymorphism analysis of mitochondrial DNA in 87 individuals. Their study revealed a clear separation between eastern (Florida and Georgia) and western (principally Alabama) populations. Each population was clearly defined by the existence of characteristic DNA sequences, implying long-standing genetic separation. Such distinctiveness is the result of the animal's habit of remaining close to the natal nest, and rarely if ever dispersing further afield. A decade later, Avise and his colleagues published similar data for the red-winged blackbird, the first such study of a geographically widespread bird species. A sample of 127 individuals from across North America revealed a genetic uniformity in the mitochondrial DNA, with only small differences between populations.

Although almost 90 percent of red-winged blackbirds nest within 80 miles of their hatching site, the remainder travel much farther, sometimes as far as 700 miles. Surprisingly little gene flow among populations is sufficient to maintain genetic similarity between them: just one or two individuals need leave their native population per generation. The redwings' dispersal behavior, albeit limited, therefore explains the lack of geographic population structure at the genetic level. By contrast, geographical populations differ significantly in their morphology, apparently in response to local ecological influences, and 23 subspecies have been recognized.

By now, phylogeographic studies have been carried out on more than 100 species of vertebrates, invertebrates, and plants. The correlation between a species' dispersal ability and its genetic structure across the landscape is strong, as in the cases of the pocket gopher and the red-winged blackbirds. Also strong is the effect of longer-term historical factors. For instance, the oscillation of glacial and interglacial periods during the Pleistocene (2 million to 10,000 years ago) drove constant changes in ecosystems: forests fragmented and migrated toward the equator during frigid times, forming smaller ecosystems called refugia; in times of more equable climate, forest refugia expanded, coalesced, and migrated away from the equator. Such changes are likely to be reflected in genetic substructuring of populations, both within and between species, depending on how species were able to respond.

One interesting such case concerns European oaks, particularly the indigenous species *Quercus robur* and *Q. petraea*, which reveal an ancient east-west geographical division in their continentwide distribution. G. M. Hewitt and his colleagues at the University of East Anglia, England, selected an unusual genetic marker to look for signs of deep history: an intron in a transfer RNA gene taken from a chloroplast. Certain regions of this intron have been highly conserved for a billion years, remaining identical in all photosynthetic organisms. Hewitt and his colleagues sequenced this region in half a dozen species of oak, making two surprising observations. First, both *Q. robur* and *Q. petraea* contain an identical point mutation, the mutation of a T nucleotide to a C in the same position. Second, the mutation is present only in the eastern populations of both species.

These two species, in company with other tree species, survived glacial periods in refugia in southern Europe. Hewitt and his colleagues suggest that the unusual mutation arose in one of the oak species during one of these ancient periods, and was later transferred to the other species by interspecific hybridization, a common process. During warmer times both species migrated north, but still clearly maintained their east-west divide. Hewitt's group points out that this same geographical divide, detectable only at the genetic level in the two oak species, forms a boundary separating taxa of many kinds, discernible in their separate morphologies. The

The distribution of oak tree species in Europe shows a clear east-west divide, apparently influenced by an ancient geographical barrier. This tree is *Querus robur* and is growing in England.

same line divides migration paths for many bird species, some taking a westerly route over Spain and Gibraltar while others go east, over the Balkans and the Bosphorus. A major geographical barrier is clearly at work here, and has been for a very long time.

A similar interplay between geology and climate must have occurred in many parts of the world, and molecular insights into the process are beginning to accumulate. One example comes from a montane forest region of northeastern Australia. The forest exists as two main areas—the Daintree forest to the north and the Atherton Tableland forest to the south—joined by a narrow corridor. During glacial periods the forests shrank in area, and the joining corridor, known as the Black Mountain Barrier, vanished. The periodic isolation and rejoining of the major forest regions should have created genetic differences between populations of all species living in the two regions (depending on their ability to disperse). Evolutionary ecologists have long debated the importance of this pattern in generating biological diversity. Some ecologists believe that during periods when species are fragmented into geographically isolated populations, sufficient genetic difference can become established between them to lead to the evolution of new species. In other words, the periodic formation of refugia promotes the evolution of new species and thus enhances species diversity.

Craig Moritz and his colleagues at the University of Queensland recently reported a study of six species in these forests, undertaken partly as a test of the debated issue. Four of the species were endemic (localized, native species)—one lizard and three birds. These were a species of skink, the grey-headed robin, the chowchilla, and the Atherton scrubwren. The two nonendemic species were the yellow-throated scrubwren and the large-billed scrubwren. Skin from 41 skinks and blood from 102 birds provided DNA for sequencing a short stretch of the cytochrome b gene taken from the mitochondrial genome. In all but one of the species, there was a clear division between the northern and southern populations, implying a concordance of geographical histories.

Nevertheless, there was considerable variation in the degree of difference: the sequence divergence was as high as 8 percent for the skink and as low as 0.1 percent for the large-billed scrubwren. Species with good dispersal abilities will have genetically similar populations because of interbreeding, as predicted by phylogeographic principles. But another factor is the timing of the onset of genetic divergence within each species: populations began diverging several million years ago for the skink and less than a million years ago for most of the birds. Overall, the results support the view that refugia are important in generating biological diversity.

The Atherton scrubwren was the exception to this pattern; its two populations displayed little variation between them. This is particularly puzzling because the bird's current distribution is restricted to several higher-altitude rainforests, which would have become cut off from one another during glacial periods. A possible explanation, suggest Moritz and his colleagues, is that during the last glacial peak, 18,000 years ago, the bird's habitat disappeared entirely from one of the forested regions. When warmer conditions returned 10,000 years ago, individuals from the surviving populations might have recolonized the other forest. If this is correct, then a similar genetic pattern should be evident in other species that share the Atherton scrubwren's habitat needs.

The final example in this section again involves concordant patterns across several species, this time in the southeastern United States—an ecologically and geographically complex region. The importance of this work, which has been carried out principally by John Avise and his colleagues for more than a decade, lies in its clear demonstration that contemporary genetic structure in a group of unrelated species can reveal the influence of historical factors—such as climate change—on population distribution, even though the precise nature of those factors may remain unknown. Furthermore, in this case the genetic structure that exists today would not be predicted based on prevailing ecological selection pressures, thus emphasizing the importance of history.

The Florida peninsula protrudes southward into subtropical waters, creating Atlantic and Gulf of Mexico ecosystems that share many species but are unique in others. The dozen or so glacial periods of the Pleistocene had a range of climatic and related effects, including a fall in sea level by as much as 300 feet and the prevalence of drier, cooler conditions. The combined influence of these changes would have been complex. Avise believes, however, that their impact on population distribution in the past can be discerned in the genetic structure of modern populations, like a historical footprint.

Avise and his colleagues analyzed patterns of mitochondrial DNA types in 19 freshwater, coastal, and marine species from this region (including various fish, eels, oysters, terrapins, crabs, and birds), producing a biogeographic history of the fauna. Most strikingly, they found a geographical concordance of genetic patterns in the populations of most, but not all, species. In marine and coastal species there is a divide between Atlantic and Gulf populations. A similar east-west division is present in most freshwater species, too. Data that have been published more recently from several other laboratories yield a similar east-west distribution pattern for several terrestrial species, including rodents, tortoises, and birds. These results imply that physical factors, generated by climate during the Pleistocene, divided species of all kinds into eastern and western populations. Thus separated, the populations gradually developed distinctive genetic characteristics that are detectable today, in the mitochondrial DNA and elsewhere.

This strong east-west population distribution pattern would not be predicted from modern ecology. To the human eye, no obvious biogeographic division can be discerned as one traverses from east to west. Indeed, the north-south temperature gradient of the Florida peninsula and the general pattern of river drainage would favor a north-south division. With so complex a climatic and ecological history to untangle, it is not yet possible to say exactly how the historical biogeographical dynamic translates so strongly to the modern genetic pattern. Nevertheless, says Avise, these surprising results "add a historical perspective to population structure and intraspecific evolutionary processes."

The extensive (and often deep) genetic differences between populations of the same species revealed by this work have implications for conservation biology and population management. Conservation biologists sometimes propose bringing together two populations of the same endangered species. This may happen, for instance, because one population has become too small to sustain itself, or because a local habitat is in danger of being destroyed. Mixing individuals from two populations under such circumstances may seem a reasonable action, but carries the danger of mixing populations that, even though they are of the same species, are of distinct genetic makeup. Furthermore, we shall see in the next section that the process of identifying species that might be endangered is made more problematical by the existence of genetically distinct populations.

Population Bottlenecks

There is a growing realization among ecologists that species are being driven to extinction at an accelerating rate, principally through the destruction of their habitats by human agency. The inexorable logging of rainforests, for example, may cause half the world's species to become extinct before the middle of the twenty-first century. As a species' habitat is eroded, its survival is threatened—a threat that becomes overwhelming even before the last member dies, when the population is reduced to numbers so small that it becomes vulnerable to inimical events such as disease, fire, and hurricanes. Small populations may also be susceptible to various genetic problems, as we shall see. Conservationists are therefore urgently directing their efforts at restricting human-caused habitat destruction and establishing protected areas for threatened species.

Ecologists have recently begun to exploit the new techniques of molecular biology to underpin such efforts with scientific data. This new field of conservation genetics has two principal goals: first, to understand the genetic changes that affect survival of threatened species; second, to use genetic information in the successful management of threatened species. The insights from these two lines of research have not always been what might have been expected.

We saw in Chapter 4 that most natural populations support a significant degree of genetic variation, a discovery that caused considerable surprise in the mid-1960s. Initially, population geneticists focused on how much random variation a population could tolerate without negative effects. Later, the question was turned to the adaptive benefit of extensive variation, an issue that remains unsettled.

The conservation community became alerted to such discussions in the mid-1970s when a study of northern elephant seals off the west coast of North America showed no genetic variation in 24 proteins. This unusual state of affairs was the outcome of voracious hunting during the last century that had reduced the population to a handful of individuals. The species survived, partly through legal protection enacted in 1922, and the popula-

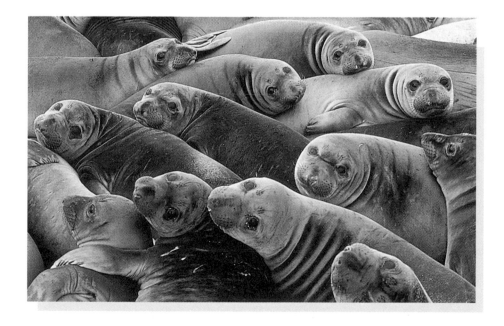

During the last century, excessive hunting by humans reduced the population of northern elephant seals to a handful of individuals. Since they began to be protected in 1922, their numbers have swelled to 120,000.

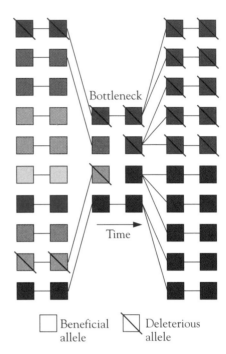

Beneficial allele □   Deleterious allele ◨

Populations that fall to a low number of individuals and subsequently recover are said to have gone through a population bottleneck. Genetic variation is typically reduced during the process, and often deleterious variants that were at low frequency in the parent population become more common, reducing the fitness of the recovered population.

tion has since grown to some 120,000 individuals. When a species passes through such a population bottleneck, much of the genetic variation distributed across the population is lost stochastically, especially when recovery is slow. Yet, because the species did not become extinct as a result of this severe population bottleneck, the significance of the loss of genetic variation was not apparent.

The true dangers of reduction in population size, subsequent inbreeding, and loss of genetic variation were illustrated by an investigation in 1979 of 24 captive populations of wildlife species. Katherine Ralls and her colleagues at the National Zoo, Washington, reported that in almost every case, infant mortality increased when inbreeding took place. Conservation biologists realized that such inimical effects—which also include reduced sperm count, reduced sperm viability, and vulnerability to disease through a reduction of genetic variation in the immune apparatus—could affect natural populations that are reduced in size. The detrimental effects of inbreeding occur through a combination of the loss of beneficial alleles and the expression of rare deleterious alleles that are otherwise swamped in a large population.

Since these early studies, biologists have investigated dozens of wild species that have passed through population bottlenecks, examining, variously, protein polymorphism, mitochondrial DNA variation, and DNA fingerprint information. Whether the original cause of population reduction was hunting, habitat destruction, or natural causes, such as disease epidemics, the danger that viability would be reduced was large, but not ubiquitous. Some species, such as the Florida panther and the cheetah, suffered severe detrimental effects, while others, such as Swiss mice and Channel Island foxes, apparently incurred no ill effects. This discrepancy plays into the debate over the adaptive benefits of genetic variability within populations. But, as Stephen O'Brien has suggested, whether a population suffers genetic problems following a bottleneck is determined as much by the *quality* of genetic variants that remain as by the *quantity* of that variation. Some populations may be lucky as they lose genetic variation through the bottleneck, in that the genes they retain, although low in number, are highly beneficial, not deleterious.

The older established techniques of traditional population genetics—such as looking for protein polymorphism with gel electrophoresis—can give important insights into a population's genetic status, that is, the extent of genetic variation present. But the newer techniques of molecular biology give even clearer information and they give it in greater quantities, through direct access to the DNA. These techniques can also be used to determine *when* in a species' history the population bottleneck occurred. Modern conservation genetics can therefore help to reveal a species' history, its present status, and its future prospects.

An extreme case of population collapse, with extreme genetic consequences, is provided by the Florida panther. This animal is a subspecies of puma that, a century ago, occupied habitats across much of the southeastern United States. Today fewer than 30 animals survive in the wild, all in the Florida Everglades. Hunting and road kills were once the main threat to the Florida panther's survival, but now the species faces an internal danger: its impoverished genetic package. Not only is the males' sperm count significantly reduced, but 95 percent of the sperm in each ejaculate are malformed. A rare heritable defect that causes one or both testicles to remain undescended has risen from 0 to 80 percent in the past 15 years. A congenital heart defect has begun to appear, and the animals' parasite load is enormous. Mitochondrial DNA analysis and DNA fingerprinting reveal

The Florida panther is extremely endangered in the wild and suffers from genetic defects resulting from low population numbers.

The Cheetah Controversy

Biologists generally agree that there is a link between the extent of genetic variation in a species and its fitness, understood in terms of reproductive success and potential for future evolution. Even though the precise nature of the link remains to be determined, low genetic variation tends to reduce fertility and increase postnatal mortality. Genetic variation has therefore become an important focus in conservation genetics, particularly in attempts to preserve endangered species. Principally through the work of Stephen O'Brien and his colleagues at the National Cancer Institute, Maryland, beginning in the early 1980s, the cheetah has become the symbol of an endangered species whose present plight and uncertain future are the result of drastically lowered genetic variation. In recent years, however, this work has come under increasing criticism, both for O'Brien's claim that the cheetah's genetic variation is unusually low for a carnivore, and for his conclusion that the species' plight is genetically caused. Differences of opinion are wide and have been strongly expressed, so that scientific papers frequently include "the cheetah controversy" as part of their titles.

Although they once enjoyed a virtual worldwide range, cheetahs are now confined to sub-Saharan Africa and a small pocket in northern Iran. Between 1960 and 1974, the African population was cut in half, to about 15,000, largely as a result of natural habitat being lost to agricultural land and farmers killing the animals to prevent loss of livestock. Although few firm figures are available, the population is thought to have continued its drastic decline. The species is now listed as endangered, and efforts to breed cheetahs in captivity have been increased, as perhaps the only hope of saving the species from extinction. Although the cheetah has relatively large litters for a large carnivore species, with an average of 3.5 cubs, only about 5 percent of the animals born reach maturity. Captive breeding has always been difficult for the cheetah, with its low rate of fertility (around 15 percent of mature animals caught in the wild reproduce in captivity) and high rate of juvenile mortality (greater than 30 percent by six months of age). Concern for the cheetah's future survival is therefore based on reality.

In 1981 O'Brien and his colleagues were invited to study the animals at the DeWildt Cheetah Breeding and Research Center, in South Africa, in the hope that they could discover a reason for the poor reproductive success there. They found that sperm concentrations were one-tenth those in the domestic cat, and that levels of abnormal spermatozoa were exceptionally high, 71 percent as compared with 29 percent in the domestic cat. (The same level is seen in wild-living individuals.) Using electrophoresis, they then examined 52 proteins from a population of 50 animals, and found no variation, or polymorphism, in any of them. Typically, 10 to 60 percent of a species' genes are polymorphic. A more extensive survey of protein polymorphism (using two-dimensional electrophoresis on 155 proteins) in southern African animals turned up just 3 percent polymorphism, or about a third of that seen in human populations. Such genetic uniformity is characteristic of an inbred population.

Three more lines of evidence seemed to support the conclusion that genetic variability was unusually low. The first was a measure known as fluctuating asymmetry. Anatomical features that are normally mirror images of each other, such as the left and right sides of the skull, become less symmetrical as a result of inbreeding. A study of the skulls of various cat species found among museum specimens apparently indicated greater asymmetry in cheetahs. The second was the fact that unrelated individuals were able to tolerate skin grafts for much longer than is normal. This implies a low genetic diversity in the major histocompatibility complex (MHC), a group of proteins that flag individual identity on the surface of cells. The third was the deaths in 1983 of half the cheetahs at the Wildlife Safari Park in

Cheetah populations in the wild have extremely low genetic variation, apparently the result of historical bottlenecks. The principal threat to their survival, however, is the loss of habitat.

(continued on following page)

(*continued from previous page*)

Oregon from feline infectious peritonitis, even though, for instance, none of the 10 African lions at the zoo developed symptoms. The cheetahs seemed much more vulnerable to this pathogen than would be expected or normal. Such extreme susceptibility to disease is consistent with a low genetic diversity in the species' immune system, which is therefore unable to mount an effective attack against infection.

All these lines of evidence led to an "emerging profile of a species in unusual genetic peril," O'Brien and two colleagues wrote in 1986. They went on to conclude that the lack of genetic variation was the result of at least one and probably several demographic bottlenecks in the species' history, during which the population dropped to very small numbers. During such events, the genetic variation that exists in the species can be eroded, by the process of stochastic loss of polymorphisms. (The bottleneck population has to be extremely small and the recovery extremely slow for high genetic diversity to be reduced to low diversity, however.) O'Brien and his colleagues concluded that the cheetah's near genetic uniformity is a significant factor in the species' decline in the wild, and in the disappointing results from captive breeding.

During the last several years, however, a growing number of studies has thrown doubt on this interpretation. Although the extent of genetic diversity is probably unusually low, as O'Brien and his colleagues have maintained, it does not seem to contribute to the species' present predicament. For instance, Tim Caro, of the University of California at Davis, and Karen Laurenson, at the University of Stirling, Scotland, have documented that 73 percent of cub deaths in the Serengeti are the result of predation by lions and spotted hyenas; most of the remainder are caused by maternal abandonment or accident, and only a few percent are the result of genetic causes. Similarly, the great majority of infant deaths in captive breeding programs are unrelated to genetic defects. The conclusion that predation, not genetic impoverishment, is the major cause of early death in the wild is supported by the presence of a booming cheetah population in Namibia, where lions and hyenas have been shot to near local extinction by farmers protecting their livestock.

Nadja Wielebnowski, a colleague of Caro's, has determined that the success rate with the

that the Florida panther has the lowest genetic variability of any puma subspecies in North or South America.

One plan to save the Florida panther from immediate extinction is to introduce members of a closely related subspecies, captive bred in Texas. The plan is controversial, however, and illustrates the extreme dilemmas conservationists often face. The goal would be to boost the population's genetic variability, thus perhaps reversing the inimical effects of inbreeding.

cheetah in captive breeding programs is very similar to that with other cats, such as the snow leopard, cervil, tiger, and lion, despite the cheetah's lower genetic variability. And Donald Lindburg and his colleagues at the San Diego Zoo have found that the high level of sperm abnormalities is no barrier to achieving high rates of fertilization. One of the main impediments to breeding success in zoos is the very artificiality of the surroundings and its effect on breeding behavior. In the wild, cheetahs are virtually solitary animals, with males and females living separately. The female advertises her readiness to mate at estrus by releasing pheromones that males can detect at great distance, and certainly out of sight. In captive breeding programs, zoo managers often waited for the female to go into estrus, and then introduced the male into her enclosure, usually with disappointing results. Lindburg and his colleagues discovered that if a male is put into an estrus female's enclosure in her absence for a short time, before bringing her back, thus mimicking in a modest way what happens in the wild, the two animals mate almost 100 percent of the time.

Just recently, the eminent Oxford ecologist Robert May has stepped into the debate over the importance of genetic versus other factors in the cheetah's plight in the wild and in captivity. His conclusion: O'Brien wins, on points. Of the contention that the cheetah's genetic variability is unusually low, May says, "I think O'Brien's case is persuasive." Nevertheless, he adds, its impact on captive breeding is probably less important than better management practices. Score one for the critics. The bottom line in May's opinion, however, is that low genetic diversity is important for the species' future, particularly in relation to susceptibility to disease. Noting that the maintenance of habitat is the prime concern of conservationists, he concludes that genetic diversity "remains an important consideration for many conservation programs, and particularly for cheetahs."

An important shift is therefore in the making. Genetic factors no longer hold center stage for conservationists; ecology and management practices are now recognized as important, too. O'Brien contends that he never claimed that genetics held the entire answer, and he agrees with May that habitat loss "is the primary concern for the species' future."

Success would mean, however, that in effect the Florida panther would be no more, at least as a genetically distinct population. The loss of such distinctiveness is, perhaps, a price worth paying to prevent the complete loss of the population.

A now classic example of the consequences of a population bottleneck in a wild population is the cheetah, which O'Brien and his colleagues have studied extensively in the past decade. (It has also become a matter of controversy of late: see the box on pages 154–157.) Once present in North America, Europe, Asia, and Africa, populations are restricted today to East and southern Africa. An impressive creature on many counts (it is the world's fastest land animal, for example), the cheetah has been subject to captive breeding efforts for decades, with miserable results. Moreover, mortality among those infants that were produced in captivity was alarmingly high: greater than 30 percent. Better management regimes have been producing better results in recent years.

In the early 1980s O'Brien and several colleagues concluded that the probable causes for these failures were a low sperm count and a high (70 percent) level of sperm abnormalities. Using traditional studies of protein polymorphism, restriction fragment length polymorphism of certain nuclear genes and mitochondrial DNA, and minisatellite analysis, they went on to demonstrate that the degree of genetic variation in wild and captive populations was vanishingly small, comparable to that of deliberately inbred strains of laboratory mice. The loss of variation was 90 to 99 percent. As a consequence, unrelated animals were able to tolerate skin grafts, something that is possible only in individuals that are extremely similar genetically.

O'Brien and his colleagues concluded that the species had undergone at least one (and probably several) severe population bottlenecks in its history. Exploiting the high rate at which mitochondrial DNA mutates, they were able to determine the genetic distance among living individuals in this genome, and calculated that the bottleneck occurred 10,000 years ago. This coincides with the end of the Pleistocene glaciation. Hunting in more recent times further reduced genetic variability in the shrinking population. Sadly, the cheetah is not alone in its predicament.

What Should We Conserve?

Modern techniques of molecular biology are also relevant to conservation, specifically in identifying what should be conserved. The apparently obvious answer to this—that is, endangered species—is less obvious and more complex than might be imagined. Ecologists use the unromantic phrase "unit of conservation" to denote a species, subspecies, or population that has been identified as requiring conservation.

Modern taxonomy is largely founded on the collection and classification efforts of nineteenth-century naturalists, who based their judgments on anatomical characteristics. As we saw in Chapter 2, morphology is not always a reliable guide for distinguishing differences between species and, more particularly, within species. A unit of conservation therefore may be a species, but it may be an individual population of a species, as we saw with the Florida panther. How do we know whether a species or a population needs protection? Because the legal framework for conservation is based on taxonomy, modern conservation genetics offers a way to help to get the answer right. Two examples of species listed as endangered illustrate the potential problems to be faced here. One concerns the colonial pocket gopher of North America, the second the tuatara of New Zealand.

The pocket gopher (*Geomys colonus*), which lived in Camden County, Georgia, was described as a distinct species in 1898. Many decades passed before biologists lavished any more interest on this population, which was rediscovered in the 1960s. Limited at that point to about 100 individuals, G. *colonus* was considered to be in immediate danger of extinction, and therefore was afforded protection by the state of Georgia.

There the matter rested until the early 1980s, when Avise and his colleagues conducted a series of investigations into this and neighboring populations of another pocket gopher, G. *pinetis*. They employed, among other techniques, restriction fragment length polymorphism of mitochondrial DNA. The results were unequivocal: the G. *colonus* population was genetically indistinguishable from its G. *pinetis* neighbors. Indeed, the results also revealed considerable genetic variation between eastern and western subpopulations of G. *pinetis*. "Either [the description of G. *colonus*] in 1898 was inappropriate," Avise noted recently, "or an original valid G. *colonus* species had gone extinct early in this century and been replaced by recent G. *pinetis* immigrants into Camden County." Whichever was the case, the Camden County population clearly did not merit recognition as a distinct species; therefore it did not deserve special protection.

The opposite fate attends the tuatara, an iguanalike lizard that, among other things, is distinguished by the possession of a third eye in the center of its head. Three species of this lizard were recognized in the nineteenth century, one of which subsequently became extinct. The two remaining were *Sphenodon guntheri* and *S. punctatus*. Legal protection was extended to the tuatara in 1895. The taxonomy of the lizard was subsequently reversed on morphological grounds over a period of several decades, ultimately lead-

Sphenodon punctatus, one of two species of tuatara that live in New Zealand, victims of mistaken identification.

ing to the recognition of a single species. One worker noted that the anatomical differences between populations previously recognized as different species were no greater than can be seen in a single colony. This view was accepted. As a result the sole "species" was considered to be widespread, and the loss of 10 of 40 populations over the past century was not rated as very serious.

Just recently, biologists at the University of Wellington, New Zealand, and the University of Sydney, Australia, conducted molecular genetic analyses on individuals from 24 of the 30 islands where the tuatara is known to live. Their results not only support the original specific distinction between *S. punctatus* and *S. guntheri*, but also demonstrate that *S. guntheri* is on the verge of extinction. The species is reduced to just 300 individuals on one island. Thus, the (mistaken) management of the tuatara as a single species has put *S. guntheri* at risk of instant oblivion, should a storm destroy its only remaining population on Brother Island.

These cautionary tales have different outcomes but the same message: sound taxonomy is the basis of sound population management and conservation. The new conservation genetics therefore has an important role to play in the future of the science, not as the sole arbiter of true taxonomic status but as a powerful complement to traditional approaches. As Avise has recently warned, the danger of "molecular chauvinism" in conservation genetics must be avoided, and this same advice applies to all other sectors of molecular ecology. Molecular biological techniques are unquestionably powerful in this and other realms of traditional biological investigations, but they should not be regarded as the only reliable tool to do the job.

A *human skeleton, right, in the company of relations: a*
gorilla, left, and an orangutan, middle.

Molecular Anthropology

*T*he term molecular anthropology was coined a little more than three decades ago by Emile Zuckerkandl, who, with Linus Pauling, invented the notion of using molecular evidence to uncover evolutionary histories. The occasion was a scientific meeting with the title "Classification and Human Evolution" at Burg Wartenstein, in Austria, under the auspices of the Wenner-Gren Foundation for Anthropological Research. The gathering included the major names in evolutionary biology and physical anthropology of the day, including George Gaylord Simpson, Ernst Mayr, Theodosius Dobzhansky, Louis Leakey, and Sherwood Washburn. Zuckerkandl was pushing hard his concept that molecules can be as revealing about evolutionary relationships as the more traditional anatomical evidence—in some cases perhaps even more so. The giants of the traditional approach were interested, but skeptical. At the end of it all,

Dobzhansky said to Zuckerkandl, "Maybe in twenty years time you will be able to say, 'I was right.'"

Dobzhansky's prediction was very close. By the mid-1980s molecular approaches to inferring evolutionary events in human prehistory had gained a foothold in anthropology and were poised to advance further. For instance, David Pilbeam, an anthropologist at Harvard University, wrote the following in 1984: "It is now clear that the molecular record can tell more about [ape and human] branching patterns than the fossil record does." Although most anthropologists would not go as far as Pilbeam in conceding technical ascendancy to molecular data, the discipline was beginning to accept its utility for infering phylogenetic patterns for different periods of human prehistory. Within a decade many anthropology departments in universities and museums included facilities for molecular studies, which allowed a new line of investigation into virtually every major aspect of the evolutionary history of humans.

So far, molecular evidence has had a major impact on our understanding of three aspects of human prehistory. In each case, the message from the molecular data ran contrary to prevailing wisdom and, in two of them, eventually overturned it. In the third, the jury is still out. Perhaps morphological evidence would have brought anthropologists to these new positions in due course. But there is no doubt that, of all the areas of evolutionary interpretation touched by molecular techniques, anthropology in particular has experienced an extremely important impact.

The first aspect of molecular anthropology we will consider relates to the beginning of the human (hominid) family—that is, its evolutionary divergence from some kind of apelike ancestor. The second concerns the most recent evolutionary event in human history, the origin of modern humans, *Homo sapiens*. The third offers a new insight into the peopling of the Americas. Here the molecular evidence has interleaved with linguistic data in an unexpected and remarkable manner.

In all three cases, molecular evidence yields not only estimates for the timing of events but also their pattern, and patterns in biology are the richest form of knowledge. The molecular evidence also emphasizes how tenuous sometimes is the coupling between genetic and anatomical change, as we saw in Chapter 2. The evolutionary relationships among humans and our closest relatives, the African apes, may disclose a particularly striking example of this phenomenon.

Ape Affinities

Chapter 2 has already told the story of how molecular evidence provided the first indication that our human ancestors diverged from the apes much later than had been thought—close to 5 million years ago instead of the 15 to 30 million that anthropologists had deduced based on fossil evidence. Until about 1984, however, the various kinds of molecular data produced by many different laboratories were, with very few exceptions, unable to show how the three lineages—human, chimpanzee, gorilla—might have diverged from each other. In other words, the tree had a single trunk that split into three branches between 4 and 8 million years ago, the three lineages arising at the same evolutionary moment. Such a three-way split is, of course, within the bounds of possibility, and a number of scholars felt persuaded by the evidence that this had indeed occurred. A three-way split, however, though possible, is a highly unlikely evolutionary pattern. Much more likely is that one lineage diverged first from the common ancestor, followed later by a two-way split. Most observers expected that, one day, the molecular data would be powerful enough to reveal this pattern.

The full and certain expectation was that the human lineage would be the first to depart from the common ancestor, leaving chimpanzees and gorillas as each other's closest relatives. Anyone who has seen the African apes at a zoo will have noticed how very similar they are in overall anatomy and, especially, in their mode of locomotion, known as knuckle walking. Detailed anatomical comparisons of humans and apes, using the latest, most rigorous methods of analysis, have strongly supported this intuitively obvious pattern.

From 1984 onward the weight of molecular evidence began to shift, showing more and more that, as predicted, there had not been a three-way split, but rather a divergence of one lineage, followed by a two-way split. Contrary to most observers' expectations, this newly perceived pattern allied humans and chimpanzees as closest relatives, and placed the gorilla at a greater genetic distance. In that year, Charles Sibley and Jon Ahlquist, then of Yale University, reported the data they had obtained from a DNA-DNA hybridization comparison of humans and great apes, which added strongly to this emerging picture.

The specific result was a major surprise. They concluded that the first hominoid to diverge was the gorilla, some 8 to 10 million years ago, followed by a divergence between chimpanzees and humans 6.3 to 7.7 million years ago. (Despite certain technical problems with this experimental system, another independent group later confirmed this conclusion using the same technique.) Incredulity is the simplest description of the reaction to these results—for good reason: every anatomical analysis had concluded the opposite configuration.

In the decade after 1984 more than a dozen analyses of hominoid evolution were published, based on different kinds of molecular data. Some used chromosome structure, others used restriction enzyme mapping, or protein electrophoresis, or aspects of mitochondrial DNA data, or nuclear DNA data of various kinds. Although there was no consensus, the weight of evidence followed the direction established by Sibley and Ahlquist. A recent review of these analyses counted 13 in support of a chimpanzee-human link, compared with 7 in support of a chimpanzee-gorilla association. At the level of DNA sequence comparison, 9 of 10 such studies support the human-chimpanzee link.

Morris Goodman and his colleagues recently pooled these sequence data and counted the number of nucleotide positions that favor one association or the other. Thirteen datasets on various nuclear genes covered a total of 37.1 kb (kilobases, or thousand bases) of DNA, with sequences from the globin gene cluster forming more than half. Within this pool of nuclear DNA, 62 nucleotide positions support the human-chimpanzee association and 25 the chimpanzee-gorilla association, while 16 favor humans and gorillas as closest relatives. The pool of sequences from mitochondrial DNA is smaller (6.6 kb), but the message is the same: 75 nucleotide positions for human-chimpanzee, 52 for chimpanzee-gorilla, and 37 for human-gorilla. One interpretation of these data is that not only are humans and chimpanzees each other's closest relatives, but also the time between the branching off of the gorilla and the subsequent human-chimpanzee dichotomy was very short.

Some molecular anthropologists disagree with this view. They point out that although the data from some genes indicate a short time between the gorilla's divergence and the subsequent human-chimp split, other data imply a long time. However, Jeffrey Rogers, of the Southwest Foundation for Biomedical Research in San Antonio, Texas, argues that this pattern can be explained if the three lineages diverged essentially simultaneously, but carried with them different gene variants. A thought experiment will illustrate his argument. Imagine that an ancestral species possessed a gene

Morris Goodman, a pioneer in the development of molecular anthropology, seen here in a photograph taken in the early 1970s, a decade after he proposed his first reclassification of humans based on molecular evidence.

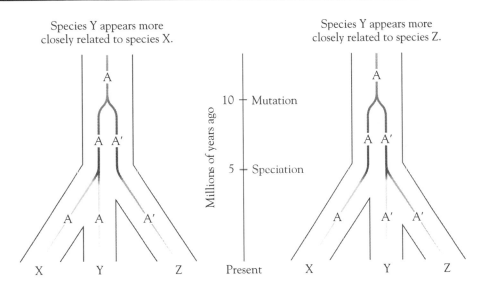

Species Y appears more closely related to species X.

Species Y appears more closely related to species Z.

10 — Mutation

5 — Speciation

Millions of years ago

Present

The generation of gene variants (polymorphism), and their subsequent distribution when new species are formed, can influence phylogenetic conclusions. Here, for instance, a comparison of the A gene variants in species X and Y would imply that they diverged 10 million years ago, not 5 million years ago, as actually happened. With the distribution of the A gene variants shown on the left, species X and Y would appear to be more closely related to each other than either is to species Z, whereas in fact they are all equally related to each other. With the distribution as shown on the right, species Y and Z would appear to be each other's closest relatives. This phenomenon has been suggested to explain the relationship among humans and African apes, as inferred from molecular data.

A. Now imagine that a variant of the gene, A', arose, say, 10 million years ago; the gene is now polymorphic. Individuals in the population of the common ancestor may now have two copies of variant A, two copies of variant A', or one copy of each variant. Suppose, finally, that 5 million years ago the ancestral species split into three daughter species, X, Y, and Z. In the population that leads to X, the variant A' is lost, leaving just A. In the population that leads to Z, variant A is lost, leaving just A'. (We'll come to daughter species Y in a minute.)

The first thing to note here is that if we compared the sequences of this gene in species X and Z, they would indicate that the species diverged 10 million years ago, despite the fact that the speciation event occurred only 5 million years ago. This erroneous dating, based on the conflation of

In Search of the Missing Link

In his 1871 book *The Descent of Man*, Charles Darwin predicted that the fossil remains of the earliest human ancestors would be found in Africa, because our closest living relatives, the African great apes, live there today. When in 1925, however, the Australian anatomist Raymond Dart announced his discovery of what he judged to be an early human ancestor, anthropologists were skeptical—not least because the specimen had been unearthed in South Africa. Moreover, the fossil, which Dart named *Australopithecus africanus* (southern ape from Africa) was far *too* apelike in appearance for most anthropologists' taste. Although the creature appeared to have walked upright as humans do, it had an ape-sized brain and a protruding face. At that time the cradle of mankind was considered to be Asia, not Africa, and our ancestor was expected to be more noble than a mere ape.

Almost a quarter of a century was to pass before Dart's claim was accepted as valid and *Australopithecus africanus* was viewed as humankind's most primitive known form—ancestor of other species of *Australopithecus* (all of which became extinct) and the *Homo* lineage, which eventually gave rise to *Homo sapiens*. Although it was impossible to place an accurate date on Dart's fossil because of the geological context in which it was found (limestone cave deposits), it was estimated to be some 2 million years old.

Australopithecus africanus maintained its status as the most primitive human form until 1978, when American anthropologists Donald C. Johanson and Tim D. White named newly discovered fossils that date to between 3 and 3.75 million years ago *Australopithecus afarensis*. The fossils had been found in the Hadar region of Ethiopia and at the site of Laetoli in Tanzania. The species, which includes the famous Lucy partial skeleton, was even more apelike than A. *africanus* in the shape of its face and cranium and in its dentition. And it was considerably older.

Like A. *africanus* before it, A. *afarensis* was said to be ancestral to all later members of the human family. As with living apes, males of the species were significantly bigger than females, said Johanson and White. Not all anthropologists agreed. The different body sizes displayed in the fossil specimens indicated the presence of several species, the critics argued, not a single species with great sexual dimorphism in body size.

In addition to issues of anatomical variation, one argument in support of viewing fossils of A. *afarensis* as multiple species rested on time and on the evolutionary pattern typical of other large

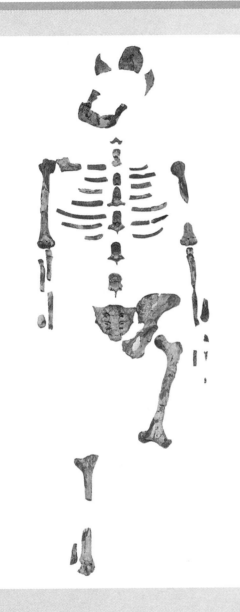

mammals. If the human family originated considerably earlier than 3.75 million years ago, as most anthropologists believed, then it was unlikely that a single species, ancestral to all subsequent species, would have been present at this later point. The usual evolutionary pattern is the radiation of many species once a new adaptation (in this case, bipedalism) has evolved, giving an evolutionary tree with the shape of a bush. Later, the bush may be pruned, leaving fewer species or, as in the case of the human family, just one.

The controversy over whether the fossils that had been assigned to A. *afarensis* represent one species or several continued for more than a decade and, for some, remains unresolved. The question of whether A. *afarensis* represents the most primitive form of the human family has, however, been settled by the most powerful testimony of all: the discovery of new fossil evidence.

Late in 1994, White and his colleagues Gen Suwa and Berhane Asfaw, Ethiopian scientists, announced their find of human fossils—even more apelike than A. *afarensis*—at Aramis, in the Middle Awash region of Ethiopia. Moreover, these fossils were dated to almost 4.5 million

The famous Lucy skeleton, left by a member of the species *Australopithecus afarensis*, discovered in Ethiopia in 1974.

(continued on page 170)

Tim White (right), Berhane Asfaw, and Gen Suwa, examining the recently discovered fossils of the new hominid species *Ardipithecus ramidus*.

years ago, close to the estimated time of divergence of the ancestor common to humans and African apes. So apelike are the Aramis specimens that some authorities wonder whether they are ancestors of the modern chimpanzee rather than of humans. But White and his colleagues point to aspects of dental anatomy and indications of bipedality in the cranium and an arm bone to support their conclusion that *Ardipithecus ramidus*, as they named the fossils, is humanlike, although its primitive nature was recognized

through the bestowal of a genus name different from the rest of the early human species. (*Ramid* is a local Ethiopian word for root, and its use as a species name here makes an obvious claim.)

The very close similarity of A. *ramidus* to apelike anatomy is to be expected in very early humans (if indeed A. *ramidus* is human). The common ancestor of apes and humans would undoubtedly have been apelike — but not like modern apes, for they, too, have evolved. Earlier than 5 million years ago, many ancient ape species are known, though there is no consensus as to which might be ancestral to later hominoids (that is, modern apes and humans). It may seem safe to say that the very earliest human species would have been apelike in virtually every respect except in its mode of locomotion—that is, in walking upright. Unless, of course, the *common* ancestor of apes and humans was a biped, with humans retaining this primitive condition and chimpanzees and gorillas developing a more specialized form of locomotion (knuckle walking). Geologists working with White and his colleagues showed that the ancient environment in which A. *ramidus* lived was not the savannah landscape in which popular mythology places the earliest humans. Rather, it was woodland and forest edge, typical of modern chimpanzee habitats.

The mid-1990s have been a fruitful time for discovery of fossils of early humans. In addition to A. *ramidus* in Ethiopia, fossils of slightly younger age have been found in northern Kenya by Meave Leakey. They are different anatomically from A. *ramidus* and A. *afarensis* and have been given a new species name, A. *anamensis*. And in May 1996, French researchers discovered part of a lower jaw of yet another new hominid species, A. *bahrelghazalia*, which lived between 3 and 3.5 million years ago. These discoveries make the evolutionary radiation of early human prehistory bushier than any anthropologist had imagined.

In one of the most surprising discoveries of all, David Pilbeam of Harvard University and Yves Coppens of the Collège de France reported the discovery of a jaw, belonging to what they judge to be a species of *Australopithecus*, in Chad. The specimen, between 3 and 4 million years old, is anatomically different from A. *afarensis*, again encouraging the speculation that a plethora of species existed at that time in human prehistory. The most remarkable aspect of this new fossil, which has yet to be assigned a species name, is not its age or its distinctive anatomy, but its geographical location.

It has become a matter of received wisdom among anthropologists that human evolution occurred exclusively east of the Great Rift Valley, while west of that valley lay the land of apes. The discovery in Chad overturns this illusion. Just as has been discerned from the fossil record regarding pig evolution, early human species would have occupied suitable habitat very quickly after their appearance; merely by following localized forest fringe, they could have soon spread through Central and West Africa. In fact, there is no way to be certain whether the first human species evolved in East Africa, South Africa, or West Africa. Geologically speaking, the time between first appearance and population dispersal would be essentially instantaneous in the fossil record. Tracking the specific point of origin and subsequent spread is therefore impossible.

so-called gene trees and species trees, would be a misleading result of gene polymorphism.

What about species Y? If its population lost variant A, a comparison of all three species would imply that Y is more closely related to species Z than to X; similarly, if Y lost variant A', it would appear to be more closely related to species X. Yet we have posited that all three species are equally related.

The conclusion from this model is that, for ancestral species in which many genes are highly polymorphic, no simple, single picture will emerge from comparisons of its descendants' genes. This, suggests Rogers, explains the mixed data for the hominoids. "The most likely hypothesis concerning the phylogenetic relationships among *Homo*, *Gorilla*, and *Pan* [chimpanzees] is an effective trichotomy," he concludes—a three-way split.

If further analyses of gene sequences prove that Rogers's hypothesis is conservative and that humans and chimpanzees are indeed each other's closest relatives, then anatomists face a daunting challenge in seeking and identifying telltale anatomical clues that reveal the true relationship. The knuckle-walking behavior and associated anatomy of chimpanzees and gorillas look so specialized that the assumption of closest relationship between the two apes was eminently reasonable. Knuckle walking evolved once, this viewpoint dictated, in the common ancestor of gorillas and chimps, and both ape lineages retained it—or so it had been thought. If the counterintuitive pattern implied by the molecular evidence is correct, then a different set of evolutionary circumstances must surround the origin of knuckle walking. Either the gorilla and chimpanzee lineages evolved knuckle walking independently; or the common ancestor of humans and both African apes (before the gorilla split off) was a knuckle walker, with humans becoming the odd ape out by developing an upright, bipedal mode of locomotion. Statistically, the second option is the most likely, because the independent origin of a complex set of anatomical adaptations (in this case, to knuckle-walking anatomy) in two lineages is extremely unlikely.

Meanwhile, in 1990 Morris Goodman, who in 1962 had proposed uniting African apes and humans in the family Hominidae, with the orangutan off in its own family, once again proposed a reformulation of hominoid classification—different from his earlier one, but as radical. His scheme is as follows. Humans and all the apes, including gibbons, would be members of the same family, the Hominidae. The gibbons would then have a subfamily to themselves (Hylobatinae), with the African apes, orangutans, and humans sharing a second subfamily (Homininae). Within this subfamily,

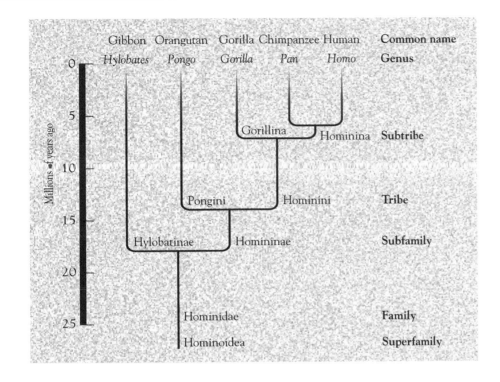

orangutans would be the single member of the tribe Pongini, while gorillas, chimpanzees, and humans would all be members of the tribe Hominini. Within this tribe, gorillas would be the single member of the subtribe Gorillina, while chimpanzees and humans would reflect their close relationship by sharing the subtribe Hominina. This classification, which is gaining support, differs from the 1962 proposal in stipulating the affinities between humans and apes as even closer than had earlier been perceived.

The Origin of Modern Humans: One Event or Many?

One of the most vigorously discussed issues in anthropology these days is the origin of anatomically modern humans. Molecular evidence has had a large influence in promoting this interest. As happened over the issue of the origin of hominids, many anthropologists vigorously rejected the first conclusions based on molecular evidence when they were presented in the late 1980s, and some still do. But the tenor of the debate can now be described as lively—not rancorous, as hitherto. Molecular anthropology has matured as a science.

The question at issue is, How did modern *Homo sapiens* arise from its immediate ancestor? In addition to fixing a time and a geographical location for this event, the molecular data also address patterns of evolution and therefore offer a richer biological insight into our origins.

Over the years, two contrasting models have been developed. Both deal with similar precursors to modern humans, but offer different patterns for their final appearance. There is general agreement, for instance, that at some point between 1 and 2 million years ago, populations of *Homo erectus* moved out of Africa and began to occupy parts of the rest of the Old World. The famous fossils of Java Man and Peking Man are descendants of these migrants. Toward about 250,000 years ago, evolutionary change is apparent in populations throughout the Old World, with the appearance of what is termed archaic sapiens, a catchall phrase that includes forms more advanced than *Homo erectus* in Africa, Asia, and Europe, and the Neanderthals, in western Asia and Europe. This change principally took the forms of an increase in the size of the brain, a reduction in the thickness of the brain case, and a diminution in the size of bony ridges above the eyes. At this point, the two models diverge.

One of them, the multiregional evolution model, argues that the process of evolutionary change that produced the archaic sapiens populations simply continued, eventually giving rise to modern humans in all populations in the Old World. There is said to have been extensive gene flow among these populations throughout all this time. If this model is correct, anthropologists would expect to find in each region of the world some local anatomical characters that are present from the very earliest to the most recent of times. In addition, the genetic roots of geographical populations would be very deep, having been established with the first arrival of *Homo erectus* into those localities.

The second model is very different. This one (sometimes called the Noah's Ark model or the Out of Africa model) views the origin of modern humans as a discrete evolutionary event, located within a single population of archaic sapiens in Africa. Descendants of this population of modern humans then spread out into the rest of the Old World, completely replacing existing archaic sapiens people. If this were the case, we should discern no general pattern of continuity of regional anatomical characters through time. The genetic roots of modern geographical populations would also be very shallow, deriving from the recent, single ancestral population. Some anthropologists suggest a less rigid version of this model in which, instead of modern humans completely replacing existing archaic populations, a degree of interbreeding occurred. Such a mixing of populations would blur both the anatomical and genetic pictures just predicted.

There exist many famous fossils from this critical time period that display aspects of anatomical change toward a modern human anatomy. Nevertheless, the fossil record is much less complete than is popularly imagined and distinctly less so than scholars would like. As yet there is no consensus over its interpretation. Some anthropologists argue that regional continuity

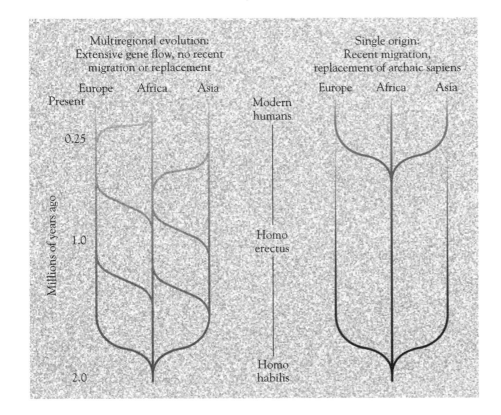

A diagrammatic representation of two models of the origin of modern humans. The multiregional evolution model argues that ancestral populations of *Homo erectus* throughout the Old World evolved in near synchrony into *Homo sapiens*. The single origin model posits that *Homo sapiens* originated in a recent speciation event in a single population, probably in Africa.

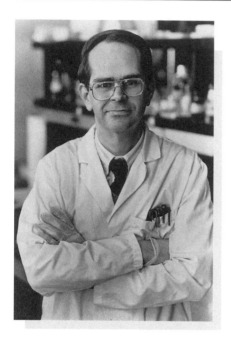

Douglas Wallace, a pioneer in
mitochondrial DNA research as
applied to human prehistory.

of anatomical characters can be detected in many parts of the Old World, thus supporting the multiregional model. For instance, the heavily built skull and facial features common both to earlier and more recent fossil humans and to modern native Australians is said to betray regional continuity in southeast Asia. The small face, small nose, and shovel-shape upper incisor teeth of northern Asians (particularly in China) have been similarly adduced in support of the multiregional evolution hypothesis. Many anthropologists point out that these features are not unique to the geographical regions in question, and argue for the absence of any evidence of regional continuity, thus supporting the Out of Africa model. Still others argue that there are indications of a single, discrete origin, but with substantial later interbreeding. About the only issue upon which there is almost (although not quite) complete agreement is that the Neanderthals became extinct and played little role in the ancestry of modern Europeans.

The problem posed by these conflicting models was well suited to the application of molecular techniques, particularly the analysis of mitochondrial DNA. Its rapid rate of evolution and maternal mode of inheritance offer a way of investigating relatively recent evolutionary history, which in this case would stretch back 2 million years at most. By now it is the best-known genome in eukaryotic cells, since its complete sequence (some 16,594 base pairs) and gene organization have been determined in humans. The application of PCR allows any chosen section of the genome to be compared between individuals. Two laboratories initially began exploiting mitochondrial DNA data more than a decade ago: they were Douglas Wallace's, then at Stanford and now at Emory University, Atlanta, and Allan Wilson's at the University of California, Berkeley.

In 1983 Wallace and his colleagues published the first survey of mitochondrial DNA in modern populations, based on a restriction enzyme mapping that effectively sampled about 9 percent of the genome. They made a series of observations that have remained valid. The first was that the total amount of variation in modern mitochondrial DNA is small and may imply an origin of modern humans placed about 200,000 years in the past. (An alternative explanation is that modern humans evolved in the ancient past, but recently passed through a population bottleneck that reduced genetic variation. There is no evidence in support of such an explanation, and considerable genetic evidence against it.)

Second, of all the populations tested, Africans displayed the greatest degree of genetic variation. Wallace and his colleagues pointed out that if all human populations accumulate mitochondrial mutations at the same rate, then Africa must be the point of origin of modern humans. In other words, assuming equivalent rates of accumulation of mutation, then the higher level of variation in the African population compared with populations from the rest of the world can be explained only by the African population's excess of another variable: time. The African population would have to be the oldest.

Third, geographic populations had developed distinctive features in their mitochondrial DNA, from which a tree could be constructed representing the history of the populations. Although Wallace and his colleagues concluded in the 1983 paper that Asia, not Africa, was the point of origin of modern humans (they asserted a faster mutation rate in Africans to explain the greater mitochondrial DNA variation among them), they later changed their interpretation in favor of Africa. That remains their conclusion today.

Four years after the publication of Wallace's 1983 paper, the notion of so-called Mitochondrial Eve became widely discussed, prompted by the publication in January 1987 of a paper by Allan Wilson and his colleagues, describing mitochondrial DNA data from 147 individuals representing Africa, Asia, Australia, Europe, and New Guinea. The technique used, as by Wallace's team, was to produce a restriction enzyme map of the mitochondrial genome. The analysis revealed 133 different types of mitochondrial DNA, which were organized into a genealogical tree by parsimony analysis. "All these mitochondrial DNAs stem from one woman who is postulated to have lived about 200,000 years ago, probably in Africa," concluded the paper. The term Mitochondrial Eve was coined in a newspaper report, and quickly became incorporated into the professional and nonprofessional literature.

Colorful though it is, the term is misleading, and led to wide misunderstanding. The reason that mitochondrial DNA types in today's human population can be traced back to a single female is not because she was the only woman living then (nor was the population she came from necessarily small), but because of the dynamics of loss of the mitochondrial DNA. This is best explained by analogy.

Imagine a population of, say, 10,000 mating pairs, each with a different family name. Now imagine that as time passes the population remains stable (each couple produces only two offspring). In each generation, on average, one quarter of the couples will have two boys, one half will have a boy

177

The famous "horseshoe" tree derived by Allan Wilson and his colleagues from restriction enzyme mapping of mitochondrial DNA taken from individuals from all geographic regions of the world. The inferred evolutionary relationships among the 182 different mitochondrial types (outer edge of tree) implies an African origin of modern humans. The greater extent of genetic variation among African populations also supports an African origin.

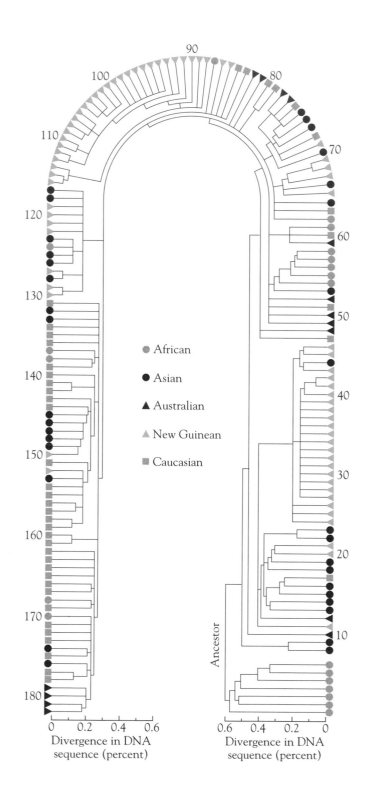

African
Asian
Australian
New Guinean
Caucasian

Divergence in DNA sequence (percent)

Divergence in DNA sequence (percent)

Ancestor

and a girl, and one quarter will have two girls. Assume that, as is common in the West, the father always passes on his family name to the offspring, while the mother never does. In the first generation, therefore, one quarter of the family names will be lost. As each generation passes, more losses will occur, although at a slower rate. After about 10,000 generations (twice the number of original females), only one name will remain. The same pattern holds for the loss of mitochondrial DNA types, except that transmission is through the female line.

Wilson and his colleagues' conclusion in the 1987 paper merely pinpointed the *region* of origin of modern humans (Africa) and gave an estimate for the *time* (about 200,000 years ago). Nevertheless, it was widely—albeit wrongly—assumed that a population bottleneck was a necessary element of the Eve hypothesis. Perhaps the Biblical nomenclature evoked a solitary female with her Adam, alone on the planet. When data from a set of genes associated with the immune system indicated that no such bottleneck had occurred in recent human history, many observers declared the hypothesis falsified. This is not the case. Eve may prove invalid, but not because recent human history experienced no population bottleneck. Eve, according to the Wilson hypothesis, would have been a member

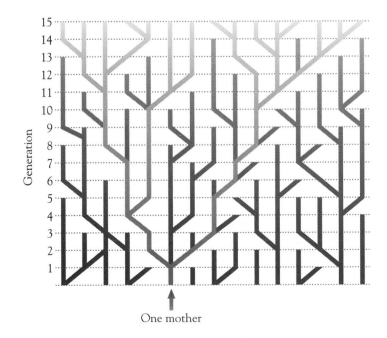

The tree demonstrates how it is possible for all the maternal mitochondrial lineages in a modern population to trace back to a single lineage in an ancestral population. At each generation, one quarter of the mothers will have two male offspring, one quarter will have two females, and one half will have one female and one male. The mitochondrial lineages of mothers that have only males will come to an end, and eventually one lineage will dominate the entire population.

179

of a large population of archaic humans, numbering about 10,000 individuals according to recent estimates.

The Berkeley team based their suggestion of an origin 200,000 years ago on their calibration of the rate of accumulation of mutants in a human population. Specifically, they measured the degree of accumulated genetic divergence within mitochondrial genomes in people of Papua New Guinea, on an island colonized close to 60,000 years ago. Working with a measure of the genetic divergence among the people, and the amount of time over which it developed, Wilson and his colleagues were able to calculate the average rate of accumulation of mutation. The figure for the divergence in sequence between two lines was 2 to 4 percent per million years, which means that for every 100 positions in the two genomes, 2 to 4 nucleotides will mutate over a million years. (Note that this is a measure of the accumulated difference between two genomes, not the rate of mutation in just one of them—a point that has been subject to some confusion.) The 2 to 4 percent divergence figure has been supported by other laboratories, using data from humans and many other mammalian species.

The first mitochondrial DNA data therefore conformed to the predictions of the Out of Africa model. Those predictions are, first, that only limited variation will exist in modern mitochondrial DNA, implying a recent origin, and, second, that the African population will display the most variation. By contrast, the predictions of the multiregional evolution model were *not* met. Those predictions are, first, that extensive genetic variation will exist, implying an ancient origin going back at least a million years and possibly two, and, second, that no population will show significantly more variation than any other.

The absence of any ancient mitochondrial DNA in Wilson's 1987 sample in itself deals a serious blow to the multiregional evolution model. If modern populations derive from a process of long regional continuity, then mitochondrial DNA samples should reflect the establishment of those local populations more than a million years ago, when populations of *Homo erectus* first left Africa and moved into the rest of the Old World. It also implies the complete replacement of existing archaic sapiens populations as modern *Homo sapiens* spread through the Old World. Initially it was possible to argue that the sample of 147 was too small, so that older mitochondrial DNA had been missed. By now more than 5000 individuals have been analyzed, however, and none has mitochondrial DNA with more accumulated variation (mitochondrial DNA that is older) than the first sample.

Opponents of the Mitochondrial Eve hypothesis, the most prominent among whom is Milford Wolpoff of the University of Michigan in Ann Arbor, argued initially that the rate of divergence had been miscalculated. A slower rate was correct, they said, in which case the modern human origin would move back in time. (Wilson would have had to have miscalculated by a factor of *five*, it should be noted, to give the million-year figure that the multiregional evolution model demands.) Wolpoff later argued

Milford Wolpoff, a leading proponent of the multiregional evolution model of the origin of modern humans.

181

that the rate was irrelevant—because, he said, the natural loss of mito-chondrial DNA over many generations precludes accurate reconstruction of population histories.

Wolpoff's argument is that the information available in extant mito-chondrial DNA types is insufficient to build a genealogical tree. Because some types of mitochondrial DNA are inevitably missing, having been lost through time, any tree that is reconstructed will necessarily give too recent a common ancestor. This argument is invalid, however, because extant types of mitochondrial DNA contain in their sequences a record of their relationship to each other, through the common ancestor. It is also reason-able to point out that, compared with the genetic record, the fossil record (on which the multiregional proponents lay stress) is more likely to suffer from missing evidence.

Mitochondrial DNA evidence will lead to valid inferences about popu-lation history only if the loss of types through time proceeds stochastically (at random). If for some reason that loss is seriously biased, then any infer-ences made from modern populations will be weakened or perhaps invali-dated. One potential source of bias is selection.

If, once modern humans had become established, a new mitochondrial DNA variant arose that was strongly favored by natural selection, then, given sufficient time and gene flow among populations, it could replace ex-isting variants. The age inferred for the mitochondrial common ancestor would then be artificially young. Although no example of such a pattern has been seen in other species, a theoretical argument can be made of its likelihood in humans. For instance, as Goodman has observed, most mito-chondrial DNA codes for genes rather than being noncoding. Coding DNA (that which directs the production of proteins) is a target of natural selection. Further, several mitochondrial genes produce proteins that are important in metabolic pathways that harness energy for the cells' future use. A critical aspect of later human evolution is the expansion of the size of the brain, particularly the neocortex. This organ consumes a dispropor-tionate amount of the body's energy budget, amounting to some 20 percent despite representing just 2 percent of the body's mass. Therefore, Goodman and others argue, natural selection may have favored certain mitochondrial genes relating to energy metabolism during the evolution of an enlarged brain. This remains speculation.

Alternatively, loss may become biased through vagaries of population histories, involving crashes and explosions in population size. In this case, it is just as likely that new variants would be lost as old. If modern humans really did evolve recently in Africa and then moved into the rest of the Old World, where they mated with established archaic humans, the resulting population would contain a mixture of old and new mitochondrial DNA. Indeed, there would be a bias toward the old because of the relatively small numbers of newcomers compared with the large established populations of archaic humans. The vagaries of population changes would in this case favor the elimination of the minority (the new types), leaving old types to predominate.

The publication of the Wilson laboratory's 1987 paper was followed by more data from Berkeley and other laboratories, plus analysis of it all by many on the sidelines. Although opinion was by no means uniform among the molecular biologists, the Mitochondrial Eve hypothesis was gaining support. In September 1991 a paper appeared in *Science* from Wilson and four other authors. The report described the sequence of part of the mitochondrial genome in 189 individuals from major geographic regions of the world. Again, the authors had used parsimony analysis to construct the most likely tree, and applied two statistical tests. The results gave powerful support to the African Eve conclusion. "Our study provides the strongest support yet for the placement of our common Mitochondrial DNA ancestor in Africa some 200,000 years ago," the authors concluded.

It turned out, however, that this parsimony analysis had been inadequate, as had the original 1987 analysis. The technique of parsimony seeks to find the tree that joins together all the observed variants by means of the minimum number of mutations. However, when more than 100 individuals are involved in a sample, and a similar number of informative sites on the genome, the number of possible trees becomes enormous. Analysts must consume vast amounts of powerful computer time to sort out which trees have the fewest mutations. For instance, the number of possible trees that can be derived from the data in the 1991 paper is an astronomical 8×10^{264}; even the number of shortest trees exceeds a billion. No amount of currently available computing power can sort through this fog of possibilities. Using the best algorithms and the fastest computers available, the most exhaustive parsimony analyses have racked up only 50,000 trees.

Researchers have to be selective in the trees they examine, and have assumed that their selections constitute a representative sample. This has proved not to be the case. For instance, the authors of the 1991 paper examined 100 trees. When two of the authors, Linda Vigilant and Mark

Anatomy and Artifacts: A Conundrum

Science draws its strength from patterns, particularly the congruence of patterns. For those anthropologists who support a recent, discrete origin of modern humans, such a congruence can be seen between the fossil and genetic evidence: both point to a speciation event in Africa close to 100,000 years ago. What of evidence of behavior? If anatomically modern humans evolved a hundred millennia ago, shouldn't we expect to find evidence of modern human behavior in the archeological record then, too? Alas, the pattern here is much less clear.

First, what do we mean by modern human behavior? For much of human prehistory, change in human behavior (as evidenced in the record of tool production and use) was extremely slow, even static for long periods of time. The earliest recognized stone implements date to about 2.5 million years ago, from Ethiopia and Kenya, and consist of small, sharp stone flakes and simple choppers. Their appearance probably coincides with the evolution of the first species of *Homo*. Known as the Oldowan industry, this primitive culture persisted essentially unchanged for a million years. Then a new form emerged, the Acheulean, again in East Africa. In addition to Oldowan-like implements, the Acheulean is characterized by teardrop-shaped hand axes, which required a more sophisticated conceptualization of shape and more manual skill in their manufacture.

Not until 250,000 years ago does the archeological record reveal another significant change, one including a new method (the Levallois) of making flakes that allowed the production of a far greater variety of implements. Virtual technological stasis again prevailed, this time terminated by the most abrupt transformation of all. For the first time, materials other than stone—such as bone and ivory—became important in tool manufacture, and the tools themselves were extremely fine and varied. Moreover, body adornment (as grave goods attest) and other forms of artistic expression appeared for the first time. Human settlements increased in size, and there is the first evidence of long-distance contacts and trade of objects, both utilitarian (stone) and nonutilitarian (shells and amber). The abruptness and magnitude of this change reflects the modern human mind at work.

Because of the way terminology developed early in the history of science, the names of these archeological periods differ in different parts of the world. In sub-Saharan Africa, the period encompassing Oldowan and Acheulean assemblages is known as the Early Stone Age. The change at 250,000 years ago ushers in the so-called Middle Stone Age, and the final transition is into the Later Stone Age. In Europe, Asia, and northern Africa the equivalent periods are known as the Lower Paleolithic, the Middle Paleolithic, and the Upper Paleolithic, respectively. (The Lower Paleolithic begins later than the Early Stone Age, because human ancestors did not expand into Eurasia until close to 2 million years ago.)

Our question here is this: When did the Middle-to-Upper Paleolithic and Middle-to-Later Stone Age transitions occur? If this apparent signal of the appearance of the modern human mind is concurrent with the evolution of modern human anatomy, we would expect to see evidence of it first in Africa, close to 100,000 years ago, and later in Eurasia (if the Out of Africa model of modern human origin is correct).

In western Europe the Upper Paleolithic signal is first seen at about 40,000 years ago. This coincides closely with the first appearance there

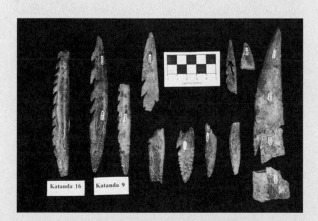

Bone tools, some of them barbed, from the Semliki River excavation, in Zaire. Until these artifacts were unearthed, the earliest known tools of this form had been found in Europe and dated to 30,000 years ago, that is, 60,000 years later than these African implements.

of anatomically modern humans, who, presumably, were migrating from the east. When we look at Asia, however, the expected congruence between anatomy and behavior quickly breaks down, presenting anthropologists with a conundrum. In the Middle East, for instance, fossil evidence for anatomically modern humans is present close to 100,000 years ago, at the cave sites of Skhul and Qafzeh. This is consistent with an origin of modern humans in sub-Saharan Africa a little earlier, and their subsequent spread northward. However, there is no sign of modern human behavior until 40,000 to 50,000 years ago.

Moreover, the Skhul and Qafzeh people apparently coexisted with Neanderthals for at least 50,000 years until the latter disappeared, presumed extinct. Throughout that period, the only known artifacts in the region are typically Middle Paleolithic in form (known in this region as Mousterian). They demonstrate no obvious behavioral difference between the archaic humans (Neanderthals) and the anatomically modern humans who eventually replaced them.

The same conundrum is seen in Africa, where the archeological picture is less clear—not least because there are fewer sites. There is no question that Later Stone Age culture (qualitatively equivalent to the Upper Paleolithic) is present in sub-Saharan Africa by 40,000 years ago. But again, if anatomy and behavior are linked, evidence of more sophisticated tool making and social structure should be present at least 60,000 years earlier. It is not; at least, not clearly so. (continued on page 186)

(continued from page 185)

There are a couple of early assemblages—the Aterian in the northwest and Howieson's Poort in the south—that include implements more sophisticated than those typical of the Middle Stone Age, but archeologists are divided over their interpretation. Do they represent glimmerings of a transition to Later Stone Age culture in a poor archeological record? Or are they merely geographical variants on the Middle Stone Age, unconnected to later, more sophisticated behaviors? There are occurrences of fine stone blades that characterize Later Stone Age industries at sites approaching 100,000 years old, but they are few—not convincing proof of the beginning of a new behavioral trend.

Early in 1995, archeologists from the United States, Canada, and Belgium reported the discovery of finely crafted bone implements at two sites above the Semliki River in Zaire. Extremely modern in form, the implements may be as old as 90,000 years, if the dating of the sites is reliable. If valid, this would be the oldest evidence of clearly modern behavior in the Old World, and it would fit the notion of an origin of modern humans in Africa. However, there is as yet no convincing archeological trail of such behavior into Eurasia.

What do archeologists make of the conundrum of an apparent uncoupling of modern human anatomy from modern human behavior? It may be that the modernization of skeletal anatomy indeed preceded the modernization of behavior, which was brought about by further biological change some time prior to 40,000 years ago. Or, as some archeologists argue, the advent

Alison Brooks, of the Smithsonian Institution, and colleagues excavate at the site of Katanda, in the Semliki River region of Zaire. The archeologists have found many tools made of bone and dated to as far back as 90,000 years ago, making them the most sophisticated tools known of that early age.

of modern behavior may have been a cultural event, not one prompted by a biological enhancement of cognitive abilities. In support of this view they cite the beginnings of agriculture 10,000 years ago, which is associated with no evidence for shifts in cognition, but rather with a modified cultural and environmental context that encouraged a new and complex form of behavior.

Stoneking, of Pennsylvania State University, were prompted to check the validity of the conclusions, they produced 10,000 trees, finding just as many that suggested a non-African as an African origin. Alan Templeton, of Washington University, and David Maddison and his colleagues at Harvard University obtained similar results.

On the basis of these data, Mitochondrial Eve, rooted in Africa, cannot be supported from a statistical point of view. Many took this setback to mean that an African origin had been disproven. But, notes Stoneking, "The fact that this conclusion is not statistically significant means only that the probability of a non-African origin is greater than 5 percent, not that there is no information regarding geographic origin." In the absence of statistically significant tree analysis, the position is as follows. First, the amount of genetic variation in human mitochondrial DNA is small and can be taken to imply a recent origin for modern humans; this conclusion is unaffected by the problems with parsimony analysis. Second, the African population displays the greatest amount of variation; this conclusion, too, is unaffected, and is most reasonably interpreted as suggesting an African origin, although without statistical weight.

Stoneking and his colleagues have recently recalibrated the rate of mutation of human mitochondrial DNA, using two new methods on data from Papua New Guinea. The two calibrations give 133,000 and 137,000 years ago as the average date of the mitochondrial DNA ancestor. Most important about the new calibration, however, is the range of dates in which one can be 95 percent confident: for one method it is 63,000 to 356,000 years ago; for the second, 63,000 to 416,000. The lower end of the range would fit well with observations about the beginnings of modern human behavior as seen in the archeological record. More significant for our purposes, however, is that even at the upper end of the range, the dates proposed are still less than half what is required by the multiregional evolution model (a minimum of one million years).

Others have analyzed the mitochondrial DNA data and come to different conclusions. For instance, Christopher Wills, a geneticist at the University of California at San Diego, looked at a subset of nucleic acid mutations that, he believes, gives a cleaner result than the total data. This analysis gives a date for the origin of modern humans somewhere between 436,000 and 806,000 years ago. This would imply that Eve was part of an earlier stage of human prehistory than many believe—that is, she belonged to *Homo erectus* rather than archaic *Homo sapiens*. But it still does not place her with the earliest migrations out of Africa (perhaps as early as 2 million years ago), as the multiregional evolution hypothesis holds.

Once imagined to be the potential answer to the question of when and where modern humans originated, evidence from mitochondrial DNA is now more properly regarded as just one line of evidence. Although there are 37 genes in the mitochondrial genome, it effectively acts as one genetic locus, because it is segregated from nuclear genes and evolves as a unit—in effect, it is like a single gene, not 37 different genes. Difficult as they are to work with, other genes are going to be important in the battery of genetic evidence.

Data from several studies of variation in nuclear genes have indicated that African populations separated first from Eurasian populations, and that this split was followed by a more recent split between Caucasian and Asian people. A similar, though more general, conclusion flows from data on a small, noncoding section of chromosome 12 produced in Ken Kidd's laboratory at Yale University. Variation in this locus is much greater in Africa than in the rest of the world. "It's very consistent with a recent spread of modern humans out of Africa," observes Kidd. "It's hard to imagine anything else that could explain the data." Similarly, Luigi Lucca Cavalli-Sforza and his colleagues at Stanford University find high variability in another kind of noncoding DNA in African populations, compared with populations elsewhere. "This observation supports the notion of an African origin," they concluded in a report in *Nature* at the beginning of 1994.

This pattern of high genetic variability within Africa and lower variability elsewhere may have been produced as follows. Modern humans arose first in Africa, allowing genetic variability to be established there. Small bands of this much larger population then expanded their territory beyond Africa and spread into Europe and Asia. These migrant bands, because of their small numbers, would have possessed just a subset of the original genetic variation of the total population. These newly established populations would then have expanded rapidly in virgin territory (as far as humans are concerned, that is), a process that would have maintained genetic variation at a relatively low level. Several research teams have looked for other examples of this genetic imprint, as an indication of the demographic pattern just described.

For instance, Wilson and Anna Di Rienzo reported such a signal in mitochondrial DNA data; it showed a burst of variation becoming established in Europe, beginning some 60,000 years ago. "One explanation is that we are seeing the consequences here of rapid population expansion that followed the migration out of Africa," suggests Di Rienzo. By contrast, Alan Templeton of Washington University finds no such signal in the data. He takes this as evidence in favor of the multiregional evolution hypothesis. In his analysis of mitochondrial DNA data, Henry Harpending of Pennsylvania State University sees evidence of separate expansions worldwide, which also supports the multiregional hypothesis. Alan Rogers of the University of Utah has applied a "sudden expansion" computer model to mitochondrial DNA data and finds evidence for the Harpending expansions, beginning about 80,000 years ago. There is no unequivocal conclusion in favor of one hypothesis or the other, notes Rogers, but the Out of Africa viewpoint receives the greater support.

It is more difficult to derive a time of origin using nuclear DNA variation, but two estimates produced in the past several years—one by Masatoshi Nei and his colleagues at Pennsylvania State University and the second by Cavalli-Sforza and his Stanford colleagues—yield numbers close to 100,000 years. This is strikingly similar to the date produced from the majority of the mitochondrial DNA analyses. Even more striking is the report of a comparison of a 729-nucleotide segment of a gene on the Y chromosome, published in mid-1995 in *Science*. The Y chromosome is the male equivalent of mitochondrial DNA, in that it is inherited in only one parental line—the paternal. Robert Dorit, of Yale University, and two colleagues found *no* variation in this sequence among 38 males from around the world. The absence of variation was a surprise and presented a challenge for scientists analyzing the history of the gene. Using statistical techniques, Dorit and his colleagues reported that their work "provides an expected date for the last common male ancestor of 270,000 years (with 95 percent confidence limits of 0 to 800,000 years)."

If the Out of Africa model is correct, then various consequences will have impressed themselves on the prehistoric record. Many of these would involve aspects of material culture, but one concerns the spread of language. It is possible that the founding population of modern humans spoke a single language, and this language would have spread with migrating populations, evolving locally as it went, just as genes do.

If the multiregional model is correct, however, then languages in different parts of the Old World will have extremely ancient and disparate roots, as with genetic heritage. There would be very small likelihood that the ge-

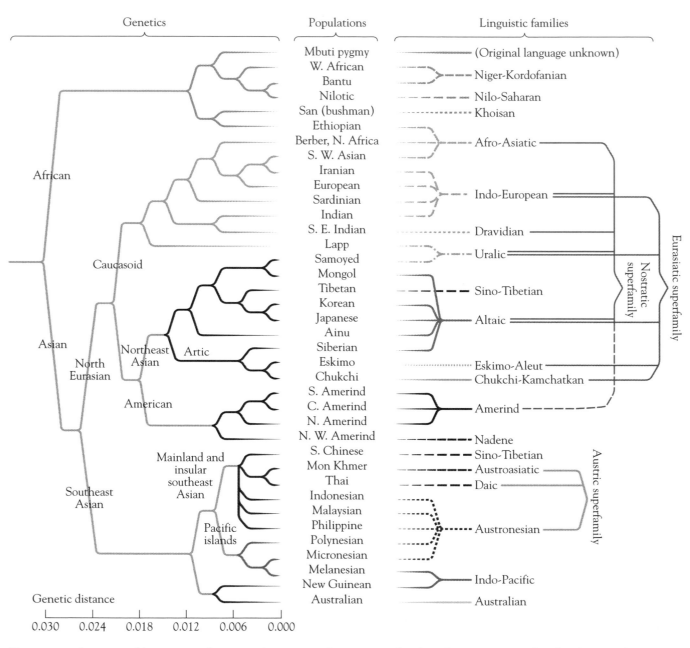

Genetics	Populations	Linguistic families

Comparison of genetic and linguistic evidence reveals a surprisingly close match among world populations: the genetic map corresponds to the map of related language families. The match encourages the notion of a relatively recent origin of modern humans, diverging from an ancestral African population.

ographical distribution of languages could be matched to specific genetic profiles among local populations. So much time would have elapsed since the origin of language and the origin of regional populations that any link between the two would have become too blurred to discern.

Although it had not been his intention to address this linguistic issue when he began amassing genetic information on human populations more than a decade ago, Cavalli-Sforza recently realized that he would be able to apply a test. Encouraged by Stanford linguists Joseph Greenberg and Merrit Ruhlen, Cavalli-Sforza and his colleagues essentially matched a genetic map of the Old World with a map of related language families. "Linguistic superfamilies show remarkable correspondence with the two major clusters [in the genetic pattern], indicating considerable parallelism between genetic and linguistic evolution," they concluded.

Although some aspects of the Stanford work are controversial, the pattern inferred from it is consistent with the notion that one of the last elements in the evolution of modern humans included an advance in language capacity. It is also consistent with a recent, discrete origin of modern humans in Africa, followed by migration into the rest of the Old World—and eventually into the New World.

Peopling of the Americas

The "discovery" of the Americas half a millennium ago was a profound shock to European intellectuals. The existence of vast, unknown continents upset established wisdom about the completeness of the world. Native Americans were equally shocked, as people from unknown shores arrived in their lands. Both groups strove to explain the other's existence, employing stories ranging from the religious to the extraterrestrial. By Thomas Jefferson's time, two centuries ago, linguistic and archeological evidence had been interpreted to imply an ancestral link between Asians and Native Americans. At some point in prehistory, it was deduced, bands of people moved out of Asia and entered the far northwest of the Americas, later to expand south. The questions that scholars began to address at this time were, first, How did this colonization occur? and, second, When did it take place?

Two centuries on, scholars are still asking the same questions. Generations of researchers have scrutinized evidence from archeology, paleontol-

A putative depiction of the arrival of Columbus in the Americas and his first encounter with Native Americans, by the sixteenth-century Flemish engraver Theodore De Bry.

ogy, and linguistics, but no clear consensus has emerged. For instance, there are three major linguistic groups in the Americas. The first two—the Eskimo-Aleut of the Arctic and the Nadene of the Northwest—are relatively homogenous. The third, Amerind, displays tremendous diversity. The Eskimo-Aleut and Nadene peoples could be imagined to be the descendants of two separate migrations. More problematical is the origin of the Amerind people. Are they the descendants of a single migration who subsequently underwent extensive linguistic differentiation? Or are they the result of multiple, independent migrations from the same genetic stock? As to the timing of such putative migrations, little consensus is to be found here, either. Some archeologists argue for a date of a little more than 11,000 years ago for first entry (so-called late entry), while others see convincing evidence that colonization began as early as 32,000 years ago (so-called early entry). (See the box on pages 194–196.)

During the mid-1980s Stanford's Joseph Greenberg developed an analysis of all Native American languages. He grouped Eskimo-Aleut and Nadene as two separate language families, or phyla, as was to be expected. His grouping of all the 600 Amerind people's languages as a single phylum with a common root was, however, controversial, principally because of their diversity. Dental evidence assembled by Christy Turner, of Arizona State University, and evidence from classical protein markers, such as blood groups, produced by Cavalli-Sforza supported this three-group analysis.

Together, the three lines of evidence led to the three-wave hypothesis: the ancestors of the Amerinds entered first, followed by those of the Nadene and, finally, those of the Eskimo-Aleut. Based on the rate at which language changes through time (glottochronology), at best a rough temporal guide, dates of 12,000-plus, 9000, and 5000 years ago were deduced for these events. The uncertainty about the Amerind date reflects the lack of precision of glottochronology beyond about 10,000 years ago.

In the realm of molecular anthropology, the relative recency of human entry into the Americas calls for the analysis of a genome with a rapid rate of mutation. Hence analysis of mitochondrial DNA variation is an appropriate tool for tackling the problem. Beginning in the mid-1980s, this approach has been applied in several laboratories, so far with conflicting results. There is consensus, however, on two points: the Asian origin of Native Americans and the genetic cohesiveness of each of the three linguistic groups, the Amerind, Nadene, and Eskimo-Aleut. The disagreement concerns the timing and pattern of entry.

The laboratory of Douglas Wallace first broached the subject with a comparison of restriction fragment length polymorphisms in the mitochondrial DNA of Pima Indians (from Arizona) and several Asian populations. They found four mitochondrial lineages among the Amerind people, which they labeled A, B, C, and D. Each was characterized by a rare Asian mitochondrial DNA marker that does not appear in European and African peoples. This result established the Asian origin of Amerinds.

The frequency of such markers in the American populations was, however, much higher than in the Asian populations. Wallace and his colleagues interpret this to imply a reduction in overall genetic variation in the New World, the result of a population bottleneck. In other words, the population that moved from Asia to the Americas to establish the Amerind group numbered no more than a few hundred or a few thousand

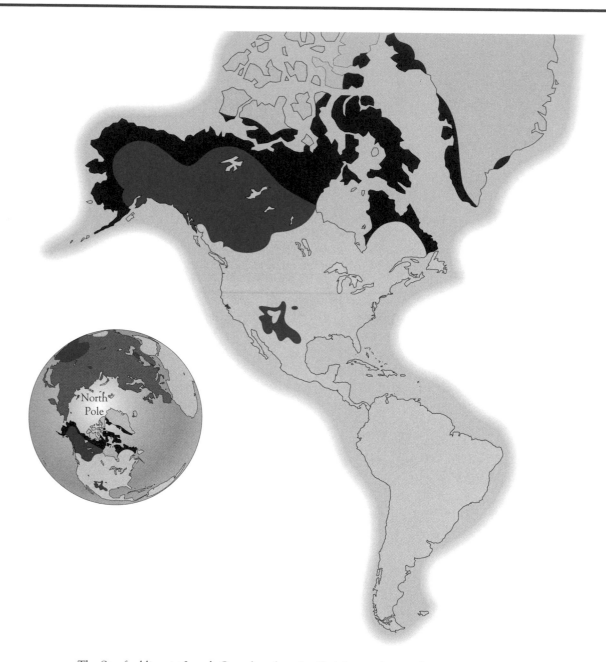

The Stanford linguist Joseph Greenberg has classified America's many languages into just three families: Eskimo-Aleut (purple), Nadene (orange), and Amerind (yellow). Amerind people were the first to enter the New World, Eskimo-Aleut the last.

and contained only a restricted range of the genetic variation present in the original population.

The Wallace group expanded the number of populations being compared to include many in South, Central, and North America (comparing Amerind and Nadene groups), plus more Asian populations. The original pattern not only held, but became more detailed. For example, the genetic variation within the four mitochondrial lineages appeared to have arisen after the putative bottleneck, strengthening the notion that a small population entered the Americas, then expanded rapidly. Moreover, while individuals in Amerind populations possessed any one of the four mitochondrial lineages (A, B, C, or D), only lineage A was present among Nadene people, indicating a separate migration. The average extent of variation within these lineages was four times greater in the Amerind than in the Nadene, indicating that the Amerind were first to arrive. Using a rate of sequence divergence of 2 to 4 percent per million years, the Wallace group calculated the time of arrival of the Amerind founders to between 21,000 and 42,000 years ago, with the Nadene arriving 5250 to 10,500 years ago. This pattern is consistent with the three-wave model, but gives an earlier entry date for the Amerind group.

More recently, Wallace and his colleagues refined their dates, using a rate of sequence divergence based on a calibration derived from genetic studies of the Chibcha-speaking people of Central America. Early in 1994, they published a revised date of entry of 22,000 to 29,000 years ago for Amerinds. (One of the four mitochondrial lineages, B, is much less variable than the rest, giving a date of 13,500 years; this raises the possibility of a second, later migration of Amerind stock.) Statistically, these results do not settle whether entry took place early or late, but they favor the former alternative.

Meanwhile, a second research effort, led by Ryk Ward at the University of Utah, has been producing different results. Initially Ward's group reported finding extensive variation in the mitochondrial lineages noted by Wallace and colleagues, which on its face implied a date of entry in excess of 70,000 years. A more reasonable interpretation, Ward and his colleagues later suggested, was that much of this variation developed in populations *before* their entry into the Americas, and that the size of the entering population was large, numbering many thousands. In this case, other evidence (such as archeological data) would be necessary to pin down the date of entry.

The First Americans: Elusive Archeological Evidence

In 1927, archeologists prospecting near Folsom, New Mexico, found spearpoints embedded in the buried skeleton of a species of bison that had become extinct at the end of the last Ice Age. This was the first important evidence for the early presence of humans in the Americas. The discovery assigned a date of at least 10,000 years ago for the peopling of the New World.

Half a dozen years later more evidence of ancient occupation came to light, again in New Mexico, which pushed the date a further 1300 years back. Once again spear points were the telltale clue to human presence, this time associated with the bones of mammoths, which paleontologists knew had become extinct earlier than the bison at Folsom. (The data on dating came later, from radiocarbon tests.) The Clovis name was applied to the characteristic form of the fluted spear points and to the people who made and used them.

During the next six decades dozens of claims were made for evidence of occupation earlier than the Clovis people, some as old as 250,000 years. Most evaporated under close examination. As David Meltzer, an archeologist at Southern Methodist University in Dallas and an expert on the peopling of the Americas, has observed: "So many false hopes have been raised about ancient sites that American archeologists have become profoundly skeptical about the new claims made almost annually." During this time a strong conviction developed that the Clovis people were indeed the first settlers, immigrants from northeastern Asia who crossed the Bering land bridge while sea levels were low as a result of glaciation. To judge from the prolific evidence of their living sites and hunting activities, these people moved swiftly from north to south, reaching Tierra del Fuego within a few centuries and perhaps causing the extinction of some 36 genera of large mammalian species as they went. Oddly, very few bones of the people themselves have been found.

The timing of entry of Asian peoples into the Americas is constrained by several factors. First, there is no evidence of human occupation of northeastern Asia earlier than 40,000 years ago; this puts an upper limit on the time of entry. Second, the fluctuating frigidity of the last glaciation would have exposed the Bering land bridge between 35,000 and 11,000 years ago, allowing passage between the continents. However, at the height of the later part of this glaciation, centered on 18,000 years ago, the eastern and western ice sheets that blanketed North America would have coalesced, preventing pas-

Clovis points, the characteristic fluted spear points made by people who lived in the Americas shortly before 10,000 years ago.

sage southward. Two windows of opportunity—from 35,000 to about 28,000 and from 14,000 to 11,000 years ago—were therefore available for migration south of the ice.

During recent years the discovery of a handful of archeological sites that may be older than 11,500 years implies that pre-Clovis people might have taken advantage of these earlier opportunities. Of these, three are found to be most convincing, even to skeptics. One is in North America, the others in South America.

The Meadowcroft rock shelter, near Pittsburgh, Pennsylvania has been under excavation for more than 20 years, and it has yielded stone artifacts on several different living surfaces, the oldest of which has been dated by radiocarbon techniques to almost 17,000 years ago. James Adovasio of the University of Pittsburgh, leader of the research effort, has faced and, for the most part, countered criticisms about possible problems with the pre-Clovis date. Life at Meadowcroft must have been chilly at that earliest date, as the southern boundary of the Laurentide ice sheet was barely 100 kilometers to the north.

In South America, the strongest pre-Clovis contender is the site of Monte Verde in southern Chile, where the remains of a series of rectangular huts has been preserved under a peat bog. Tom Dillehay and his colleagues at the University of Kentucky excavated the site between 1976 and 1985, finding hearths, animal bones and hides, and tools of wood and stone. Radiocarbon dates on charcoal in the hearths put the oldest occupation time at 13,000 years ago. A nearby site is dated—much less securely—at 33,000 years ago.

A claim for a still older date is made for the site of Pedra Furada in northeast Brazil. A rock
(continued on page 198)

(continued from page 197)

Sea levels dropped drastically during the last ice age, exposing a new coastline (green area), and the Bering land bridge between North America and Asia, enabling the first human occupation of the New World by migrants from the Old World. The timing of the peopling of the Americas is still disputed.

shelter poised above a valley floor, the Pedra Furada is decorated with more than a thousand painted signs and figures, including humans, lizards, armadillos, and jaguars. Excavation at the site, led by Brazilian archeologist Nième Guidon, has uncovered hearths and stone tools. Charcoal in the hearths was dated by radiocarbon at 42,000 years old. Guidon considers this a true reflection of the site's age, while other archeolo-

gists believe it is an overstatement. Pre-Clovis, yes, but by more than 30,000 years? Probably not, they contend.

If people were south of the ice in the Americas in pre-Clovis times, there should be evidence of early entry north of the ice, too, where game was plentiful. Two sites in the Yukon, Old Crow Basin and Bluefish Caves, have yielded flaked stone and fragmented bone that may be genuine artifacts, not products of natural breakage. Ages for these sites, which are difficult to date accurately, extend to 25,000 years ago and more.

The current archeological evidence poses many questions. What is the significance of the explosive settlement implied by this extensive record of Clovis people? Does it represent a major migration and spread of people, at such a late date? Or was it a new cultural expression that spread rapidly among people already in the New World continents? The pre-Clovis archeological record implies an earlier migration (or migrations), but its sparsity suggests limited population expansion and settlement. In any case, genetic evidence from Native Americans implies that they are descendants of people who arrived in the Americas long before the Clovis period.

By the time Columbus arrived in the fifteenth century, the Americas were home to more than a thousand languages and associated cultures. Today only half of these languages survive, and many are endangered. All bear witness to a rapid diversification of people who descended from perhaps just a few groups of migrants from the continent of Asia.

In subsequent analyses, based on a revised rate of mitochondrial DNA sequence divergence, the Utah group has proposed a date of entry located some 13,000 years ago for Amerinds and 6200 years ago for Nadene peoples. This work therefore is consistent with the late-entry model. The different conclusions offered by the Utah and Emory laboratories flow principally from different conclusions about the rate of sequence divergence. So far there is no resolution of these differences, and the timing of the peopling of the Americas remains to be resolved.

Ward and his colleagues have made observations about linguistic and genetic differentiation among the Amerind and Nadene peoples. The question they addressed was whether changes in languages and genes proceed at similar rates. To affirm that they do, genetic differentiation among Amerinds should be less than that between Amerinds and Nadene speakers, because Amerind languages form one phylum and Nadene a separate phylum.

In fact, genetic differentiation among Amerinds and between Amerinds and Nadene people is rather similar, indicating that language evolution can proceed more quickly than genetic evolution. "We speculate that this is because linguistic diversity is generated in a fundamentally different way from genetic diversity," Ward and his colleagues wrote recently. "If language change is viewed as a cultural phenomenon that is driven by social and historical events, it is likely to exhibit 'bursts' of change as well as periods of 'stasis.' By contrast, molecular evolution proceeds at a more even pace." Linguistic and genetic change will seem congruent in pattern only in populations that diverged in the distant past, allowing time for differences in the products of the two mechanisms to even out.

Molecular anthropology is now a fully established science. It does not replace traditional studies of human prehistory, of course, but is a powerful complement to them. The case studies described here reveal that power, but they also demonstrate the complexities of molecular techniques. If there were no such complexities, unequivocal answers would emerge quickly—and, as is obvious with the question of the peopling of the Americas, this has not been the case.

The quagga, a close relative of the zebra, became extinct in the late nineteenth century. It was the first creature from which ancient DNA was recovered.

Ancient DNA

*I*n this final chapter the story of molecular evolution at once moves boldly into the past and uncertainly into the future. Steven Spielberg's spectacular 1993 film *Jurassic Park* brought into the public arena the notion—more properly, the fantasy—that it might be possible to recover dinosaur DNA from the belly of an insect trapped in amber, and from it reincarnate these extinct terrible lizards. The film was based on Michael Crichton's 1990 novel of the same name, but the idea goes back a decade earlier, to the musings of Charles Pellagrino, a paleobiologist and writer at Rockville Center, New York. In 1977 he saw a piece of 95-million-year-old New Jersey amber, inside which was an apparently perfectly preserved

Chapter

8

fly, and it triggered a speculation. "Three more decades of technological advance and we may be able to extract and read DNA from flies' stomachs, where, if we are lucky, we will find the blood and skin of dinosaurs," he wrote in the science–science fiction magazine *Omni*. If there were pieces of genetic code missing, he suggested, it might be possible to reconstruct them by genetic extrapolation and by borrowing from the genes of modern dinosaur relatives. "Then everything that goes into building a dinosaur could be published in the form of chromosomes," he continued. "We could insert these into a cell nucleus, provide a yolk and an eggshell, and hatch our own dinosaur."

The three decades have not yet passed, and, *Jurassic Park* notwithstanding, dinosaurs remain in the realm of extinction. Nevertheless, the essence of Pellagrino's speculation—extraction of DNA from extinct creatures—has been fulfilled and with it has been born a new science of many parts, one that has no proper name but can be described collectively as ancient DNA research. The science was born in 1984 and, as has so often been the case in the development of studies in molecular evolution, its midwife was Allan Wilson.

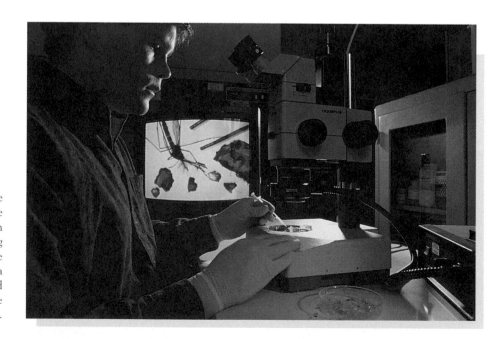

Amber has remarkable powers of tissue preservation, so insects entombed in the golden, brittle material still retain much of their cell structure, including fragments of DNA molecules. Here we see insect tissue being removed from a piece of cracked amber; the insect and tweezer ends are displayed on the video screen.

The technical challenges that need to be met in this research are twofold. First is the poor physical state of DNA in tissue that has been dead for hundreds, thousands, or even millions of years. Second is the development of a means by which such DNA can be recovered. In their pioneering work, Wilson and his colleagues took the first steps in overcoming these challenges.

With the ability to extract ancient DNA from all kinds of organisms effectively comes the opportunity to stretch back into the past the lines of research described earlier in this book. Those lines of research—into population biology, natural history, and anthropology—now yield information about the past based on genetic information in living organisms. Ancient DNA research addresses similar questions, but draws on genetic information that is part of that past. This new temporal dimension allows scientists to address issues of genetic history directly, and also enhances the power of questions that may be asked of the present, by permitting a fresh set of comparisons with the past.

Resurrecting DNA

By the early 1980s, biologists knew that dead tissue, such as animal skins, bone, and mummified bodies, contained macromolecules like protein and DNA. Indeed, several workers had extracted protein from such specimens, with the idea of obtaining amino acid sequences that would shed light on evolutionary questions. The results were not particularly promising—not least because the protein molecules were often extensively degraded. An alternative approach exploited the ability of antibodies to recognize different proteins, whether fragmented or not. This approach proved to be more successful, but it was a further step removed from the information that resided in the genes themselves, thus limiting the detail of genetic comparisons that could be generated. If genes could be recovered, even in a fragmented state, it would be possible to tap into a vastly greater pool of genetic information.

Working with Russell Higuchi in the Berkeley laboratory, Wilson decided to see whether DNA could indeed be extracted from long-dead organisms. They obtained dried muscle and skin from a quagga (*Equus quagga*) that had died in an Amsterdam zoo in 1883 and was stored in the Museum of Natural History in Mainz, Germany. That had been the last

203

remaining individual of a species that, based on anatomy, was judged to be closely related to the zebra and more distantly to the horse. Like the zebra, the quagga was striped, but only on the front half of its body.

Using conventional techniques of the time, Wilson and his colleagues obtained DNA from the quagga tissue, although only one-hundredth as much as would have been expected from living tissue. They targeted mitochondrial DNA, because each cell provides many copies of each mitochondrial genome. Again using conventional techniques, the Berkeley team selected sections of DNA and cloned them in order to have sufficient quantities of material for analyzing the nucleotide sequence. Cloning involves joining the chosen DNA fragment to a carrier DNA molecule that has the ability to replicate itself in bacteria. The technique works well with fresh tissue, but is much more problematic when used with the degraded DNA of long-dead tissue.

The first observation was that, as expected, the quagga DNA existed in short fragments, none longer than 500 base pairs and most closer to 100. (In contrast, chains of more than 10,000 base pairs are routinely extracted from fresh tissue.) The researchers sequenced and compared two short stretches of quagga and mountain zebra (*Equus zebra*) DNA containing approximately 115 base pairs each. Initially, Wilson and his colleagues counted 12 differences between quagga and mountain zebra DNA in the 229 nucleotide positions they were able to compare. Later they discovered that two of the differences were the result of the degradation process in the dead tissue. The net total of 10 differences between the quagga and the zebra implies close genetic relationship, as expected from the anatomical similarities.

In subsequent experiments, the Berkeley researchers compared these sequences with equivalent DNA from the plains zebra (*Equus burchelli*) and the domestic horse (*Equus caballus*). The tracing of the evolutionary history of the horse family from eohippus (genus *Hyracotherium*), which lived 50 million years ago, to the modern genus, *Equus*, is a legendary story in paleontology. Nevertheless, uncertainties still exist about relationships within *Equus*, particularly over the taxonomic affinities of the quagga. In recent times three main schemes were proposed, based on comparisons of various aspects of anatomy. The first grouped the quagga most closely with the domestic horse, separate from the plains and mountain zebras. In the second, the quagga was said to be more closely related to the plains zebra than to the mountain species. And the third concluded that the quagga is not a separate species at all, but merely an extreme variant of the plains zebra.

Wilson and his colleagues' molecular analysis supported this last scheme, because it showed essentially no differences in DNA sequence between the quagga and the plains zebra. Buoyed by this discovery, workers at the Vrolijkheid Breeding Centre in Cape Province, South Africa, have recently been trying to reestablish the quagga, by selective breeding of plains zebra specimens whose hindquarters' markings are slight. By the early 1990s, foals were being born that bear an uncanny resemblance to preserved specimens of quagga, in their rich, dark brown background coloring, striped forequarters, but only poorly developed stripes on the hindquarters. Hardly the stuff of Crichtonesque science fiction, the quagga nevertheless may one day walk the plains of Africa again, with ancient-DNA techniques having played an important role.

In their November 1984 paper in *Nature* describing the initial work on the quagga, Wilson and his colleagues stated that "the present report seems to be the first demonstration that clonable DNA sequence information can be recovered from the remains of an extinct species." They also noted that they had already extracted DNA from a 40,000-year-old mammoth that had been preserved in the frozen steppes of Siberia, and they speculated on the possibility of retrieving DNA from insects trapped in amber, which can be many millions of years old. Wilson and his colleagues concluded the paper with a prophetic sentence: "If the long-term survival of DNA proves to be a general phenomenon, several fields including paleontology, evolutionary biology, archeology, and forensic science, may benefit."

From Mummies to Molecular "Fishing" Trips

As Wilson and his colleagues were working on tissues from the quagga, another researcher, Svante Pääbo, was thinking along similar lines. While pursuing a thesis on molecular virology at the University of Uppsala, Sweden, Pääbo learned the ease with which DNA may be extracted from the fresh tissues of many kinds of organisms, and he wondered whether the same techniques would work with dead tissue. He was interested in human mummies specifically. Pääbo sought samples from his own university's

Svante Pääbo is a pioneer in the development of ancient DNA research. Here, he is holding the foot of an Egyptian mummy similar to that which inspired his first foray into the study of ancient DNA.

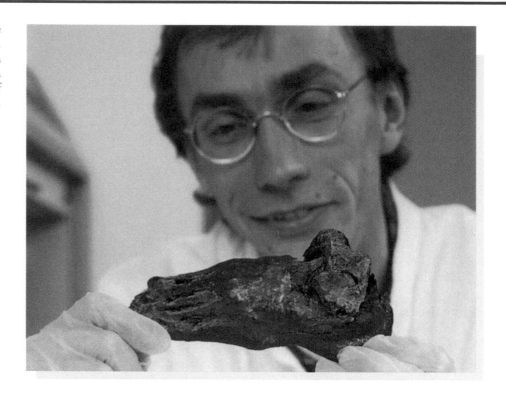

antiquities collections and from the State Museums in Berlin, where mummies are plentiful. He was able to obtain DNA from these specimens—the first time anyone had pulled genetic information from human bodies that had been dead several thousand years. But, like the Berkeley group, he found that the DNA was extremely fragmented, measuring only 100 to 200 base pairs in length.

Cloning DNA fragments of this size is often difficult, not least because the molecules are often physically and chemically modified in a way that interferes with the fidelity of replication. It is also difficult to match such short sequences with known genes, just as matching a fragment of a sentence with its place in a book is more difficult than locating a whole page. These uncertainties may raise doubts about the source of the DNA.

For instance, when Pääbo isolated and cloned a piece of DNA from an Egyptian mummy some 2000 years old, he had to be certain that it was truly human—not merely the genetic material of bacteria or fungi growing or preserved in the desiccated human tissue. This potential for contamination, always likely to be a problem with ancient-DNA research, was recognized as such from the beginning. The potential for contamination would become even more exaggerated later on. For Pääbo, the solution was to search for certain short sequences, known as Alu-repeats, that are characteristic of human genomes and are present in very large numbers. Finding and identifying them would be relatively easy.

By late 1984 Pääbo was confident he had isolated DNA fragments, some quite long, from several mummies dated between 2310 and 2550 years ago. He was unaware of the Berkeley laboratory's work and expected to be first to publish on ancient DNA. Wilson and his colleagues' November publication denied him that priority. The unwitting rivals soon became collaborators, however, and Pääbo eventually joined Wilson's laboratory in what was to be a pioneering and productive association. Before long DNA had been isolated from a handful of extinct creatures, including the ground sloth of southern Chile and the marsupial wolf of Australia. Nevertheless, the technical problems presented by the existence of ancient DNA in typically small fragments remained stubbornly persistent.

Cloning DNA is a little like molecular "fishing": you throw in the hook—the carrier DNA—and hope to pull out something interesting. You only know whether your catch is interesting after the DNA has been replicated sufficiently in a bacterium. With fresh tissue, it is possible to retrieve many different DNA fragments, producing many different kinds of clones. With dead tissue you are lucky to retrieve only a few intact fragments, largely because the bacterial enzyme systems that replicate DNA are inefficient with short fragments and often introduce errors. With fresh tissue it is possible to "go fishing" several times with a high probability of obtaining the same fragments of DNA. This allows analyses to be replicated, which is fundamental to experimental science. The very low efficiency of DNA cloning from dead tissue, however, makes repeated recovery of the same sequence a virtual impossibility, no matter how many fishing trips are repeated. Replicability was therefore beyond reach with the techniques available in 1984 and early 1985. Pääbo later described the situation as follows: "Molecular evolutionists who were keen on time travel therefore found themselves in a depressing situation. Because they could not verify their results by repeating an experiment, the study of ancient DNA could not qualify as a fully respectable science." Alec Jeffreys of the University of Leicester, England, was even more pessimistic. In a review of Wilson and his colleagues' success with the quagga, he wrote that "any hopes that molecular biology can be fused into a grand evolutionary synthesis by studying fossil DNA still look like nothing more than a glorious dream."

The position seemed hopeless, with nature holding its secrets tantalizingly out of reach. DNA is a fragile molecule, especially in the inimical environment of tissue that has just died. At death, breakdown begins on two fronts. The first is internal, the result of the tissue's own degradative enzymes. These enzymes, which can cleave many kinds of molecules, are normally packaged within organelles or other structures within the cell, and therefore never come into contact with DNA. At death, these structures begin to disintegrate, spilling the enzymes into other regions of the cell, including the nucleus. The second front is external, created by the action of bacteria and fungi, which are ubiquitous in the environment. Proteins that in the living cell offer structural support and protection to the DNA are soon destroyed by organisms of both types, leaving the DNA susceptible to enzymatic degradation. Moreover, once stripped of its protective protein, DNA is vulnerable to the effects of exposure to water and oxygen, both of which can destroy the molecule. In the longer term, background radiation also sunders DNA chains—although in most cases the long term never becomes an issue, because degradation begins so quickly after cell death. As Pääbo has put it: "The average length of . . . old molecules (about 100 base pairs) [is] the same whether the material [comes] from a 13,000-year-old ground sloth of southern Chile or a piece of dried pork just four years old." In fact, in a piece of pork just four days (even four *hours*) old, the degradation process is already well advanced. This inescapable fact of nature seemed to put the genetic past beyond scientific reach.

Success and Potential Snares

Then in 1985 Kary Mullis, a researcher then with Cetus Corporation, described his development of the polymerase chain reaction (PCR), which permits the retrieval of small fragments of DNA with precision and power. We saw in earlier chapters how PCR greatly facilitates molecular investigations in evolutionary biology, natural history, and anthropology: what had been possible, albeit difficult, in these studies previously was now much easier. The invention of PCR or something of equal power, however, was fundamental to the establishment of ancient-DNA study as a science. The fact that nature conspires to sunder DNA chains soon after death was no longer

a virtually insuperable barrier to progress; now, selected segments of DNA could be recovered from the fog of degraded and damaged molecules. The impact of PCR is a wonderful example of how advances in science may be propelled as much by new techniques as by new theories.

Wilson had close contacts with Mullis. So, joined now by Pääbo, his group was the first to apply PCR to ancient DNA. One of the first tasks was to check the data from quagga DNA, and this was the means by which they discovered the two errors in the comparison with mountain zebra DNA. Next the Berkeley group attempted to retrieve DNA from much older tissue, the brain of a human whose body had been preserved for 7000 years in the Little Salt Spring sinkhole in Florida. After working through several technical snags, the group succeeded in obtaining segments of mitochondrial DNA of a type not present in contemporary Native Americans. This was the beginning of efforts to extend into the genetic past the question of the human settlement of the Americas, as we saw in the previous chapter.

By the late 1980s and early 1990s, a dozen laboratories in the United States and Europe were enthusiastically applying PCR to tissues from many different organisms, some of which had been dead for just a few decades or centuries, others for tens of thousands, even millions, of years. In the midst of these developments, a team at the University of Oxford made an important advance. Researchers had initially judged the prospects of recovering DNA from bone as minimal, and they had therefore concentrated on soft tissues, as described earlier. In late 1987 Erika Hagelberg and her colleagues began working on new methods of DNA extraction that they hoped would circumvent the difficulties posed by the mineral environment of bone, succeeding within a year. A further year passed, however, before they could be certain that the DNA was real and not an artifact.

The report in 1989 of this confirmation was a landmark in ancient-DNA research, because it opened up the possibility, among others, of taking genetic studies of human populations deep into the past. What if, for instance, DNA could be extracted from Neanderthal bones? This would provide vital—perhaps clinching—data on the origin of modern humans. If Neanderthals were an extinct side branch of human prehistory and not a population in a continuing evolutionary lineage, then their DNA would be very different from that of later Europeans. Several laboratories are now attempting to recover Neanderthal DNA—so far without success.

The apparent success of the Hagelberg team was soon tempered by new questions, most important of which was the likelihood of contamination. Because the subject of study was a 5000-year-old human (from an archeological site in Oxfordshire), the researchers had to be certain not only that the DNA was human but also that it came from the skeleton and not from the scientists themselves. Anthropological specimens are handled frequently during research, so the prospect that old bone will become contaminated with modern human DNA is great. Just a few cells sloughed off the skin or carried in the aerosol of a sneeze are sufficient to provide the PCR technique with grist for its powerful amplification process. Bryan Sykes, a member of the Oxford team, told a cautionary tale at a recent international gathering on ancient-DNA research. His laboratory had spent some months attempting to extract DNA from mammoth bones, and eventually did so. In the end, however, what had been extracted turned out to be not genetic information from deep within the Ice Age, but modern DNA from one of the lab's technicians.

As a way of testing the fidelity of their technique, Hagelberg and her colleagues chose to extract DNA from a pig bone taken from the wreck of the *Mary Rose*, the flagship of Henry VIII that sank in the English Channel in 1545. A tasty morsel on a Renaissance dinner plate therefore became the test case in an important extension of ancient DNA research. DNA extracted from the bone was demonstrated to be porcine, not human. The

1991 success prompted a headline writer for an English newspaper to mark the event with the following waggish remark: "Pig brings home the bacon on DNA." The way was now open for molecular anthropology to practice on bones a few thousand, if not yet tens of thousands, of years old.

Mining the Museums

To others in the field of ancient-DNA research, the temporal ambitions of anthropologists were modest indeed. For a while there developed a competition to recover the oldest DNA. The challenge was effectively kicked off early in 1990 when Edward Goldenberg and Michael Clegg, of the University of California at Riverside, reported the isolation of DNA from a 17-

A parade of organisms into the past, for which claims have been made for the extraction of ancient DNA.

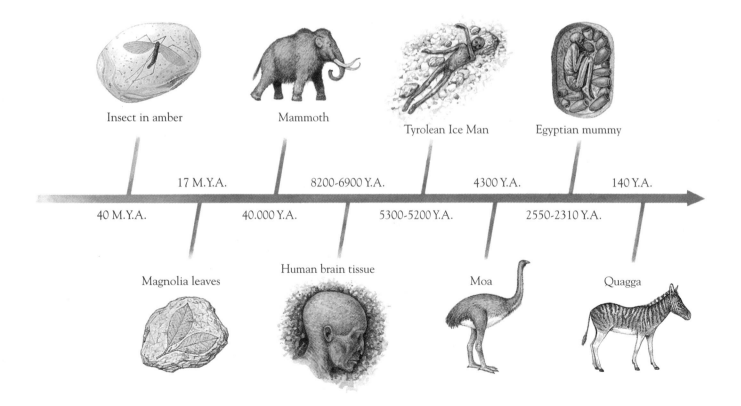

Insect in amber   Mammoth

Tyrolean Ice Man   Egyptian mummy

17 M.Y.A.   8200-6900 Y.A.   4300 Y.A.   140 Y.A.

40 M.Y.A.   40.000 Y.A.   5300-5200 Y.A.   2550-2310 Y.A.

Magnolia leaves

Human brain tissue   Moa   Quagga

211

million-year-old magnolia leaf that had been preserved in the clay beds of an ancient lake in Clarkia, Idaho. The conditions of preservation at the site must have been extraordinary, because some of the leaves were still green, indicating the persistence of chlorophyll and, by implication, other macromolecules, including DNA. The reported recovery of DNA, with some fragments as long as 800 base pairs, was greeted with mixed responses. Some workers were able to repeat the experiment, giving credence to the claim, while others, including Wilson and Pääbo, failed to do so, suggesting caution. As we shall see later, that note of caution seems to have been justified.

Before long, however, the 17-million-year date looked unimpressive as reports of, first, 25-million-year-old DNA from a termite, then 25- to 40-million-year-old DNA from a bee, and, finally, 130-million-year-old DNA from a weevil followed, the first two late in 1992, the last in June of 1993. Each of these organisms had been trapped in amber at death and, apparently, had undergone extraordinary preservation (see the box on pages 214–216). The researchers involved were at the American Museum of

An extinct termite, *Mastotermes electrodominicus*, trapped in 25-million-year-old Dominican amber, similar to the specimen that yielded the first DNA from an amber-entombed insect.

Natural History, New York, and the University of California, Berkeley. The "Million Plus Club," founded by Goldenberg, was fast recruiting new members.

There was more of importance to these spectacular announcements than simply who could lay claim to the oldest DNA in the world. Biologists having access to DNA from these organisms could address certain questions of phylogeny more effectively. For instance, the DNA recovered by the AMNH team from the fossil termite resembles DNA sequences from a living Australian termite that has been viewed as an evolutionary link between cockroaches and termites. With data necessarily restricted to a single individual, however, tests of genetic relationship within living groups and resolution of phylogenetic uncertainties among those groups depend on a tenuous thread of evidence, particularly considering the long time scales involved.

These various announcements of superold DNA inevitably received wide publicity, not least because they evoked the possibility of a real *Jurassic Park*. They also provoked vigorous exchanges among ancient-DNA researchers themselves, mostly over two issues: the value of museum collections and how best to use them; and the dangers of ancient specimens becoming contaminated with modern DNA. Neither issue was new to the field, of course, but the high visibility of the insects-in-amber work served to emphasize them.

For several centuries field biologists have assiduously collected specimens of organisms of all kinds, from butterflies to buffaloes, from berries and beetles to barnacles and birds. The back rooms of every museum of natural history are stacked high with the fruits of these efforts. New species have been named on the basis of such collections, and comparisons made among them. Memorials to both the explorers and the patient catalogers among natural historians, they long formed the foundation of traditional biology, in taxonomy and phylogeny. They were the pride of the host institutions. But with the advent of molecular biology these collections not only lost their prestige but also came to symbolize a past era of intellectual endeavor, whose time had come and gone. Indeed, several major collections have been dispersed in recent decades, to make room for the tools and subjects of modern biology.

It is paradoxical that the agent that once threatened to cast museum collections into oblivion is also the one that has now reinforced and multiplied their value. Ancient-DNA research has transformed the piles of

Still Life in Amber

The oldest forms of life preserved in amber—fungi, pollen, and flowers from coniferous plants—come from amber in southern Scotland, formed some 300 million years ago during the Carboniferous period. However, examples of amber-trapped organisms older than about 130 million years are rare. Their rarity may be an artifact of science, however, as little material has been studied in this older age range. Until a decade ago, biologists considered organisms entombed in amber to be mere curiosities—unusual objects fashioned by a quirk of nature, but of little scientific value. Long-dead organisms suspended in brittle, honey-colored tombs were regarded as shadows, empty ghosts of past lives. Then in March 1982, George Poinar and Roberta Hess of the University of California, Berkeley, published electron micrographs revealing the detailed internal structure of muscle and other cells sectioned from a female fungus gnat preserved in 40-million-year-old Baltic amber. In its slow transformation from viscous fluid to brittle solid, amber resin becomes a time capsule, preserving specimens essentially intact, into the twentieth century. DNA, albeit in fragmented form, is also preserved.

Poinar and Hess's discovery prompted the obvious question: What chemical constituents of amber resin achieve such remarkable preservation? The resins have long been used commercially as preservatives, in wine making, as local antibiotics, and in waterproofing. With such widespread application through ancient and recent history, Poinar expected that chemists would long since have analyzed the components of amber resin. Not so. To be fair, it is not completely an issue of neglect on chemistry's part, but of technical difficulty. The resins are horrendous chemical cocktails.

Amber resin is produced by many different species of Kauri pine, a relative of the monkey puzzle tree. The chemical mix includes a variety of organic molecules (sugars, alcohols, esters, and a mixture of terpenes). Resins from different tree species produce variants on this basic theme, and even the same tree may vary the cocktail. The question is, Which of these chemicals is responsible for the dehydration, tissue fixation, and exclusion of bacteria effected by resin, each of which may promote preservation? The alcohols and sugars in the resin may be efficient at extracting water from tissue, thus accomplishing the dehydration essential to preservation. Some oxidative products may generate aldehydes, which may act like glutaraldehyde, the fixative biologists use in the laboratory. However, much of the resin's preservative power may derive simply from its ability to exclude oxygen and destructive biological agents, like bacteria.

Whatever the means of preservation, it's obvious that the process must begin quickly. As soon as an organism dies, degradation of its tissues begins on two fronts: first internally, as a result of the tissue's own enzymes; then externally, by bacteria and fungi that are ubiquitous in the environment.

Matching the mystery of the resin's preservative properties is the process by which it becomes transformed into amber. The sticky, odiferous resin fairly rapidly becomes a pliable, still odiferous solid, called copal. After perhaps 4 or 5 million years, the material is no longer pliable, but forms an amorphous, nonodiferous, glasslike material—true amber. The key chemical reaction in this transformation is polymerization, whereby short terpenes become linked together to form long chains, giving the amber its solidity. Most synthetic polymers are formed by linking multiples of one type of simple chemical unit (monomer) into long chains (polymer). In amber the monomers are not all simple, and there is a mixture of them. The resulting polymer is therefore extremely complex. The physical conditions that are necessary for, or may expedite, this amber polymerization process are unknown.

Beyond these various mysteries, however, is the reality that amber preserves delicate and fragile organisms, and interactions between organisms, that conventional fossilization cannot. One example is the earliest known gilled mushroom, *Coprinites dominicana*, preserved in 35- to 40-million-year-old Dominican amber. Also preserved in Dominican amber is a bamboo seed, with a story. The top of the seed is hooked, and caught in one of the hooks was a strand of mammalian hair, which forensic examination revealed to be that of a carnivore. "Thus perhaps an early cat once brushed against a bamboo stalk while prowling through the forest and picked up

several stowaway seeds," says Poinar. "The large cat likely then scratched itself against a tree to dislodge the seeds, and one of them fell into a deposit of fresh resin." The preservation process then did the rest, delivering the specimen to scientific scrutiny 25 million years later.

Inevitably, the largest group of organisms found in amber is insects. But bacteria, fungi, and plant parts are also common. Because resin deposits are relatively small, vertebrates are rare: until early 1996 only frogs and lizards had been found; the presence of mammals and birds in the ancient ecosystem is indicated by hair (as above) and feathers. The first mammal specimen, a tiny shrewlike creature in 18- to 20-million-year-old Dominican amber, was reported by workers at the American Museum of Natural History. Even without the direct evidence of vertebrate fossils, however, their presence is testified to by the remarkable number of their preserved parasites, such as fleas, ticks, mosquitoes, biting midges, horse flies, and blood-sucking flies.

The exquisite preservation of organisms in amber means that the tiniest details of their anatomy can be compared among species, extinct and extant, giving unprecedented insight into microevolutionary changes. Amber resin also has the Pompeii-like ability to freeze life in action, at the instant when organisms became immobilized in its viscous trap. One pair of flies, in the midst of mating 40 million years ago, is to be seen in a piece of golden Baltic amber. And a piece of Lebanese amber shows a biting fly being bitten by a mite, in the oldest known example of

(continued on page 216)

215

(continued from page 215)

external parasitism. The oldest example of internal parasitism is a clearly visible nematode inside the abdomen of a midge, preserved in 135-million-year-old Lebanese amber. A similarly parasitized midge, from Dominican amber, is even more dramatic, as the nematode is to be seen emerging snakelike from the host's body. A female drosophilid fly in Dominican amber was parasitized by a different species of nematode, 120 juvenile forms of which are to be seen in the host's body, from which some are already escaping.

Another form of frozen behavior preserved in amber but rarely in other forms of fossilization is termed commensalism. Most commonly, this behavior takes the form of one organism passively carrying another: the passenger benefits but the host suffers no detriment. Many specimens of amber-entombed pseudoscorpions, for instance, are still grasping various hosts, such as beetles, wasps, and flies. Commensalism, like parasitism, would be virtually invisible in the fossil record if amber preservation did not exist. "Amber may be limited in the size of organisms that are preserved," concedes Poinar, "but its ability to freeze behavior gives us an important link between forms of life as we know them today and as they occurred in earlier times."

animal skins and trays of pinned insects or pressed plants into something far more valuable than repositories of anatomical information. They are now recognized as an invaluable reserve of genetic history—lodes to be mined in what sometimes resembles a "gene rush." Had those collectors of old not busied themselves as they did—for their own contemporary purposes—this genetic history would have been lost. True, it is still possible to find insects in amber and collect fossilized bone; but we can no longer sample populations that were alive a century ago and have since vanished.

An important concern highlighted by the insects-in-amber work is how museum collections should be exploited in the future; after all, their newfound value puts them at a new risk. Museum curators are conservative by nature and profession. Their job is to preserve specimens. Taking samples of these specimens for DNA extraction necessarily destroys them, or at least damages them bit by bit. Molecular biologists are perhaps a little imperious by nature and profession. Already, and perhaps inevitably at this stage in the evolution of the science, there have been complaints that these researchers have assumed a right of access to museum specimens, simply taking what they want without regard to the material's value in the context of the collection or the importance of the specimen itself.

The field is already maturing, however. Molecular biologists and museum biologists are coming to work together, often crossing and melding the boundaries of their previously independent disciplines, just as is happening in other areas of research in molecular evolution. A decade is a short time in the life of a science that surely has a long future, so this trend can be expected to continue. Already there is greater understanding of how

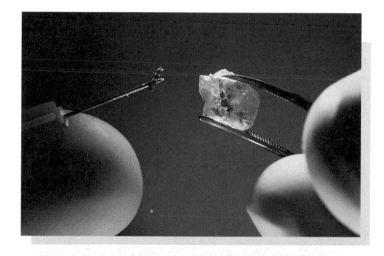

The extraction of tissue from amber-entombed insects is a delicate operation, usually carried out with fine needles and syringes under a microscope.

Samples of ancient DNA being examined under a spectrophotometer. This instrument enables researchers to visualize the DNA, which is invisible to the naked eye.

to select which specimens may be subject to sampling and who has a right to the genetic information and the amplified DNA produced in investigations. Most important, researchers have recognized the need for real biological questions to be asked, so that rare specimens are sampled not simply because of their rarity, but as part of an investigation into a larger issue of population or evolutionary biology. Irrevocable evidence, in other words, should be carefully rationed for significant projects.

The second issue highlighted by the insects-in-amber reports is that of the possibility of misleading contamination through the use of PCR. Researchers of ancient DNA had been aware of this snare. But they were forced to address it seriously when, early in 1993, the British chemist Tomas Lindahl effectively stated in a paper in *Nature* that DNA retrieved from organisms millions of years old *must* be a contaminant, because DNA simply does not survive that long.

Lindahl's credentials for so bold an assertion were powerful. More than anyone else, he has studied the fate of DNA molecules over time. In the 1970s Lindahl had performed meticulous experiments on the degradation of "naked" DNA—that is, DNA molecules in the absence of protective proteins—dissolved in water. The process was rapid and inevitable, hence his skepticism about claims for the existence of superold DNA. "It can be predicted that . . . fully hydrated DNA is spontaneously degraded to short fragments over a period of several thousand years at moderate temperatures," he wrote in his 1993 paper. Under the most favorable conditions, he continued, "it seems feasible that useful DNA sequences aged tens of thousands of years could be recovered." He deemed longer preservation to be vanishingly unlikely.

By pushing ancient-DNA researchers to examine their assumptions and experimental practices more closely, Lindahl's highly critical paper led the field in a beneficial direction. Those who plan to use PCR to extract DNA from an insect in amber, for instance, now meticulously exclude contact with researchers working with living insects, lest modern DNA contaminants be mistaken for ancient DNA. Responses to the paper also emphasized how little is known about what influences the fate of DNA in dead tissues. It is true, as many researchers pointed out, that DNA in its natural environment faces conditions very different from the conditions in water in a test tube. Whether these conditions may inhibit degradation rather than accelerate it, however, no one could say from first principles.

DNA does appear to persist for tens of thousands of years in some situations, and possibly for tens of millions of years, too—but it does so against all odds, at least as a nucleic acid chemist would see it. The mechanisms of preservation remain a chemical black box. As William Hauswirth, a biologist at the University of Florida College of Medicine, recently put it: "After ten years of experience worldwide, there appear to be no hard and fast guidelines as to what type of tissue or preservation site will yield the highest quality ancient DNA."

As with the harmonious relationship that has emerged between museum curators and molecular evolutionists, so too a spirit of collaboration has developed between chemists and ancient DNA researchers. Both are interested in discovering why DNA survives longer than it "should"—that is, what is the chemistry that allows the field of ancient-DNA research to exist at all?

Applying Ancient-DNA Research: Rats, Rabbits, and Red Wolves

This last section will offer some examples of the application of ancient-DNA research to specific scientific questions. The topics introduced are merely representative, not comprehensive, of the broad scope of research now underway. They touch on aspects of phylogeny, anthropology, and population biology. The section will finish with a return to the specter of retrieving dinosaur DNA.

The first application, in 1990, of ancient-DNA research to populations was carried out by Svante Pääbo, Kelly Thomas, and Francis Villablanca on kangaroo rats. Common in the Mojave Desert, these animals have been collected during the past century and their skins stored in museums. The Berkeley team extracted and sequenced mitochondrial DNA fragments from some 48 skins representing three populations in the Mojave region. They then collected living individuals from these same three regions, and extracted equivalent DNA fragments. Their comparison of modern and ancient DNA in the three geographical regions showed that populations of 60 to 80 years ago were very similar to modern populations, both in DNA types and variation among those types. These results indicated that the populations were stable and that little migration had taken place between the regions. This result may seem undramatic, but the study was a landmark in the development of ancient-DNA research—it illustrated how population insights could be obtained.

Humans have been influential in animal migrations for millennia, whether through accident or intention. The European rabbit is a prime example. The species (*Oryctolagus cuniculus*) originated in the Iberian peninsula and has subsequently expanded its range through much of the world, usually by means of human agency. In an attempt to understand some aspects of these population movements, a team of researchers at the Centre Génétique Moléculaire, Gif-sur-Yvette, Paris, have analyzed mitochondrial DNA patterns in modern and historical rabbit populations.

In the region of their origin there is considerable morphological and genetic diversity, with two main mitochondrial lineages: type A, in southern Spain, and type B, in northern Spain and southern France. The French team looked at archeological sites on the Island of Zembra, near Tunis, and found evidence of the introduction of rabbits in Roman times. Although the modern rabbits of Zembra resemble those of southern Spain in their morphology, their mitochondrial type is that of the northern populations. So, too, is DNA extracted from rabbit bones in Zembran archeological sites. Moreover, the degree of genetic variation in the modern populations of Zembra implies that rabbits there derive from a single, small introduction some 1400 years ago.

Ancient-DNA research is beginning to play a part in conservation efforts, because it can answer questions about recent population history that may be important in deciding the fate of apparently endangered species. For instance, the red wolf was once common in the southeastern United States, but its numbers began to decline precipitously after 1900: human incursion destroyed wolf habitat, hunters targeted the wolves themselves, and surviving members of the species began hybridization with coyotes. By the 1960s pure red wolves had virtually disappeared in the wild, except for a small population in southeastern Texas.

Two decades later the species existed only in captivity, a population built up by breeding a population of individuals captured in Texas in the 1970s. But captive-bred animals have been released in Alligator River National Wildlife refuge in eastern North Carolina to reestablish wild populations. Could the red wolf, considered by some to be ancestral to the gray wolf and the coyote, once again be safe in the wild?

There arose lingering doubts, however, about the genetic identity of the captive-bred animals. Were they really pure, or had hybridization with coyotes introduced foreign genes? Several conventional genetic analyses supported this suspicion. Robert Wayne and his colleagues at the University of California at Los Angeles joined the investigation using PCR, first on tissues from modern populations and then on tissues from pelts in museum collections. The first conclusion was that the captive-bred animals were indeed hybrids of gray wolves and coyotes. This result seemed to imply a disappointing outcome from a well-intentioned conservation effort.

But Wayne and his colleagues' further work, analyzing red-wolf skins collected in the early decades of the century, before hybridization in the wild was thought to have taken place, produced a much more surprising conclusion. These animals, too, were seen to be hybrids. A revised hypothesis became necessary—namely, that the red wolf, rather than being ancestral to the gray wolf and the coyote, was in fact a hybrid of these species, and had never been a distinct species in its own right.

We saw in Chapter 6 that conservationists often face the problem of trying to reestablish populations through captive breeding, despite the fear

The evolutionary relationship between the red wolf (pelts at right) and the grey wolf (pelts at left) was solved by the extraction of DNA from pelts collected early in this century and stored in museums. These pelts are part of the collection at the Smithsonian Institution.

that the populations they are dealing with may be genetically impoverished. Through recourse to museum collections of earlier populations of these species, they can often trace the actual history of the species' genetic variation, as has been done with the cheetah. As helpful as it is to have their hopes and fears grounded in the known genetic status of a species, however, conservationists can work only with what remains. Often, moreover, in deciding whether to breed such populations, or even to hybridize individuals from geographically and genetically distant populations, they are overwhelmed by the urgent need to do something—anything—to salvage some remnants of a genetic lineage. Information from ancient DNA research may help guide what is done in desperation, but it cannot bring back species from extinction or recover genetic variation that has been permanently lost.

Applying Ancient-DNA Research: Marsupials and Moas

Ancient-DNA research with nonhuman animals has shed light on the evolutionary history of species and genera and also on more recent population history. Two interesting cases involved the marsupial wolf of Australia and those extinct, giant birds of New Zealand, the moas.

The native mammals of Australia and South America are marsupials—in these mammals, the greater part of gestation takes place externally, in the maternal pouch. Many species are so morphologically similar between the two continents that an evolutionary relationship has been postulated for them. One example is *Thylacinus cynocephalus*, the Australian wolf, once widespread across Australia but now extinct. It has been said to be related to South American carnivorous marsupials, the borhyenids. For instance, the Australian species shares three tooth characters and a pelvic trait uniquely with the South American species; by contrast, there are just two hind-limb features in the Australian wolf that indicate it is more closely related to other Australian marsupials. This weighting of four-to-two characters in favor of a South American link has encouraged some to hypothesize an evolutionary relationship between the two, despite the fact that the two continents have been geologically separate for tens of millions of years.

222

When Richard Thomas, now at the Natural History Museum in London, and his colleagues isolated two fragments of mitochondrial DNA from Australian wolf hide, a very different picture emerged. When they compared the sequences with those of living marsupials from Australia and South America, they found *no* genetic links between the species on different continents. The Australian wolf's ancestors were Australian, not South American. The striking morphological similarities in the Australian and South American species must have been the result of strong convergent evolution. Morphologically, these marsupial carnivores are also strikingly similar to placental carnivores, also as the result of convergent evolution.

Australia's geographical neighbor, the two islands of New Zealand, was once home to a dozen genera of giant, ostrichlike birds, the moas. Few mammals existed in New Zealand in prehistoric times, and the flightless moas filled the large carnivore niche that in other continents was occupied by mammalian species. Moa populations began to decline when the first human settlers, the Maori, arrived a millennium ago, and the great birds soon became extinct. Two evolutionary questions have been unresolved concerning the history of moas: first, their origin and relation to each

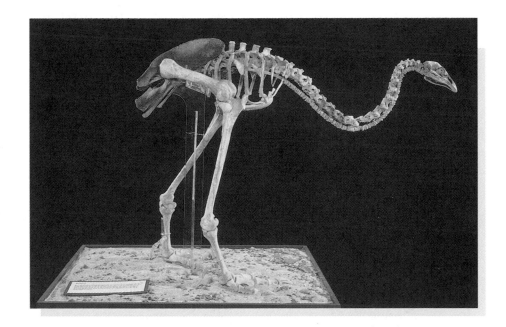

The articulated skeleton of a moa, a giant, ostrichlike bird, a dozen genera of which thrived in New Zealand prior to human colonization a thousand years ago.

other; second, their relation to kiwis, the smaller flightless birds that are a national emblem of New Zealand.

Although moas were known to have had a long evolutionary lineage (some 80 million years), mitochondrial DNA analysis of soft tissues from mummified moas showed that the group that existed when the Maori arrived traced their origins back less than 30 million years. Alan Cooper of Victoria University in Wellington, who carried out the ancient-DNA extraction and analysis, suggests that this biological history reflects the geological history of the islands. Some time earlier than 30 million years ago, the surface area of the islands was dramatically reduced to some 15 percent of its current extent, as they were plunged beneath the waves by tectonic action. Most of the island remained submerged for 7 million years, by which time most of the moa genera that existed previously had become extinct. When the land surface expanded again, about 26 million years ago, the new ecological opportunities favored rapid speciation. A new cohort of moa species and genera arose at this time.

Cooper's analysis of mitochondrial DNA also showed that all the moa genera he tested were closely related to each other—but not to kiwis, whose closest relatives are the emus and cassowaries of Australia. Very probably the kiwi's ancestor arrived in New Zealand as a flighted bird, and subsequently became flightless, as often happens on islands when predation is minimal.

Applying Ancient-DNA Research: Human Prehistory

One of the earliest and most extensive applications of ancient-DNA research to anthropology was led by William Hauswirth and his colleagues, who were able to ask questions about population structure and dynamics in an American Indian community that lived between 8200 and 6900 years ago. In the early 1980s, 177 skeletons and 91 brains were recovered from a peat bog in Windover, Florida. Unusual chemical circumstances had conspired to preserve the individuals and the DNA in their brains: the peat environment excluded oxygen, and dribbling limestone springs buffered the acidity that would otherwise have destroyed the genetic material.

The individuals present represent the passage of some 50 generations. By recovering samples of both mitochondrial DNA and nuclear DNA associated with the immune system (the major histocompatibility complex, or MHC), Hauswirth and his colleagues were able to explore whether differences existed between males and females in their social (mating) behavior. Very few mitochondrial lineages were present within the population and through the 1300 years represented at the site. One inference to be drawn is that females tended to remain in the community, with few genetically distinct females moving in from neighboring groups. By contrast, variability among the genes of the MHC was much greater than would be predicted on the basis of mitochondrial variability, implying that more marital exchanges of men than women took place between genetically distinct neighboring communities. This pattern is not unexpected for prehistoric societies, but these data represent "the first genetic evidence for such differential behavior," says Hauswirth.

The question of the pattern of settlement of the Americas can also be addressed using information from ancient DNA, although so far no settlement of the debate described in the previous chapter has emerged. If anything, some aspects of it have become more complex. For instance, the primary hypothesis of a small number of migrations is based on differences in mitochondrial DNA types among the three major linguistic groups (see page 197). According to Connie Kolman and her colleagues at the Smithsonian Tropical Research Institute in Panama, however, differences in mitochondrial types can arise easily and recently. They analyzed mitochondrial DNA in the linguistically distinct Chibcha and Choco populations in Panama, which are thought to have derived from a common population that settled the region 7000 years ago. The Choco carry all four of the major mitochondrial types found in Amerinds, while the Chibcha possess only two. This implies that mitochondrial DNA types were lost within the last 7000 years in the Chibcha, bringing into question comparisons between the patterns of DNA types in the Amerind, Nadene, and Eskimo-Aleut peoples.

There are counterarguments to this conclusion from proponents of the three-migration hypothesis, but no resolution. Perhaps it should not be surprising that as more data accumulate, the larger picture becomes more complex. The task is to extract the overall pattern from the details, as the following example demonstrates.

When Erika Hagelberg refined techniques for extracting DNA from ancient bones, one of the first anthropological questions to which she turned was the colonization of the Pacific—specifically, the history of Easter

The Ice Man

In anthropology and archeology, chance often plays a major role in exciting discoveries. So it was when, in September 1991, two German climbers came upon the corpse of a man partially embedded in a glacier at 10,500 feet in the Tyrolean Oetztaler Alps. At first, the couple thought they had found the body of a fellow climber who had recently met an unfortunate end. Such discoveries are, after all, not uncommon in this treacherous terrain, just on the Italian side of the Austrian border. But closer examination soon revealed evidence of a very different story. The dead man was dressed in crudely sewn leather garments that were lined with straw. Near him were a metal ax, a simple stone knife, a leather pouch containing a flint for lighting fires, a wooden backpack, and a bow and quiver—not exactly the accoutrements of a modern-day climber. Here, obviously, was a presence from the past—an individual who has come to be known simply as the Ice Man.

Unfortunately for science, several peremptory amateur attempts to remove the Ice Man's body from the glacier were made, thereby destroying valuable evidence. For instance, although the body was resting in a shallow trench some 60 feet long and 18 feet wide, its original posture in the trench had not been recorded, making problematic any effort to determine the cause of death. In the extraction attempts, his clothing was torn and, somewhat ghoulishly, his outer genitalia became detached when his trousers were removed. The use of a pneumatic drill during the process also damaged the left part of his pelvis. When experts finally arrived at the scene, the Ice Man was naked, apart from a right shoe. Clearly visible on his back were bold tattoos, stripes colored with charcoal.

Austrian and Italian authorities tussled for possession of the Ice Man's body, now that he was clearly recognized as a major archeological treasure. Although he had died on what is currently the Italian side of the border, Austria prevailed, because its experts were the ones to recover the body. The Ice Man's remains now rest in the anatomy department of the University of Innsbruck. He is stored at just below freezing, and researchers wishing to study him must work quickly, because after 30 minutes he must be returned to the refrigerator to prevent the deterioration that would result from thawing. Werner Platz, an anatomist at Innsbruck, directs the anthropological research, while Konrad Spindler is in charge of archeology, which includes the Ice Man's artifacts and the markings on his body. Using the degree of wear shown on his teeth as a guide, anthropologists suggest that the Ice Man lived to between 25 and 40 years of age.

At first the Ice Man was judged to have perished some 4000 years ago, because his ax was thought to be bronze and of the type found frequently at archeological sites of that age. However, the ax was subsequently determined to be made of copper, product of a softer technology that predated the Bronze Age. Radiocarbon dating of skin and bone samples, carried out independently in Oxford and Zurich, revealed that the Ice Man died between 5200 and 5300 years ago.

Skeletons of this age are common in Europe, but this is the first corpse to be found anywhere in such a high state of preservation. The Ice Man's flesh and organs are in far more pristine condition than those of desiccated Egyptian mummies or Neolithic bodies entombed in the sediments of oxygen-free bogs. Scientists are therefore afforded an unprecedented view of an individual who lived 200 generations ago, from the tattoos on his skin to the contents of his stomach. And, of course, the Ice Man arrived on the scene just as techniques for studying ancient DNA had become well developed. For these reasons the Ice Man has been described as the archeological discovery of the century.

The Ice Man lived when civilization was on the rise in Mesopotamia, during a period that saw the development of central authority, monument building, ambitious irrigation projects, and the emergence of city and nation states. Writing had been invented, in Sumeria. But for the Ice

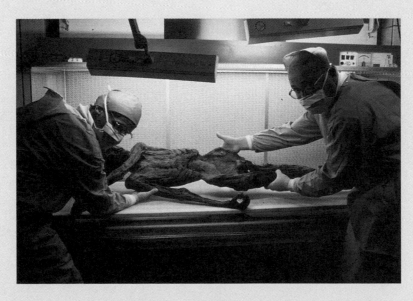

In September 1991, climbers discovered the body of a man who died in the Tyrolean Oetztaler Alps some 5000 years ago. The body, seen here, is stored under refrigerated conditions at the anatomy department of the University of Innsbruck.

(continued on page 228)

(continued from page 227)

Some of the tools that the Ice Man carried with him when he died.

Man and his fellow inhabitants of Europe, life was much simpler. They lived in small, stable villages and subsisted by farming and a measure of hunting. Equipped as he was with a bow and a quiver charged with 14 arrows, the Ice Man was ready to take down an antelope or a goat, if chance offered. The tattoos on his back, a white marble pendant on a necklace, and an indication that his earlobes were pierced for earrings speak of symbolic body adornments, which may have been similar to those developed by the earliest modern human settlers in western Europe, some 35,000 years ago.

Although he was relatively young when he died, the Ice Man was already suffering from osteoarthritis, a degenerative bone disease, in

Island. One of the most isolated inhabited islands on Earth, Easter Island has long fascinated historians, who were divided over the origin of its initial settlers. Some thought it was colonized as part of the great Polynesian migration throughout the Pacific, several thousand years ago. Others believed there had been two migrations, the first of Caucasoids from Peru, the later bringing Polynesians from the west. The most notable proponent of the latter theory was Thor Heyerdahl, who conducted heroic sea-going exhis neck, lower back, and one hip, according to radiologists at the University of Texas M. D. Anderson Cancer Center, in Houston. At some earlier point in his life, the Ice Man had also broken eight ribs and had suffered a slight fracture of one hip. In this respect, he fits a picture recently emerging from the study of ancient skeletons of just how hard life was in early farming communities. Calcium deposits in the Ice Man's blood vessels indicate the presence of heart disease—an insight possible only from the very special conditions of preservation to which the Ice Man was exposed at the time of his death.

What the Ice Man was doing in the mountains when he died is beyond speculation. Despite the fact that his body was disturbed before forensic experts reached the scene, however, it is possible to say something about the mode of his death. He had animal flesh and fruit with him, so he probably did not starve. Nor is there any indication of fatal wounds. A team of anatomists and anthropologists from Germany, Austria, and Sweden suggest that a clue may be taken from the Ice Man's left ear. He had evidently lain down on his left side, with his head on a hard surface, and gone quickly to sleep. As a result, the left ear flap was folded under his head, with no subsequent disturbance. The researchers speculate that the Ice Man was in a state of exhaustion when he reached the trench. Perhaps he sought shelter in the trench, possibly from adverse weather conditions, and lay down to rest. In the deep sleep of exhaustion, he would have been unaware of any discomfort caused by the folded ear flap. During sleep, he probably slipped into the coma that ensues when exposure drastically lowers the body's core temperature (hypothermia) and died. High winds would have dehydrated his tissues rapidly before his body became entombed in ice—an efficient freeze-drying process.

periments to see if ocean currents could take a raft from the Americas to this remote island. Similarities in material culture between Peru and Easter Island were held to support this hypothesis.

Mitochondrial DNA was removed from the skeletal remains of 12 individuals who lived on the island several hundred years ago—the evidence it provides, describing the mitochondrial DNA types present, seems to settle the issue. All 12 individuals possessed a type that Hagelberg and her colleagues have found to be characteristic of ancient and modern Polynesian populations. Known as the Polynesian type, it is marked by a nine-base-pair deletion and a three-base-pair substitution. This pattern, which is like a genetic fingerprint, marks the early Easter Island settlers as Polynesians, not Americans, as Heyerdahl believed.

Ancient-DNA research added an important footnote to the remarkable discovery, in September 1991, of a 5000-year-old mummified body buried in a Tyrolean glacier (see the box on pages 226–229). So extraordinary was the discovery that several observers suggested that it might be a fraud—that unknown perpetrators had placed an already mummified body from Egypt or South America in the recently melting ice. A large team of researchers from several European countries recently reported a DNA analysis from the body—this was no small task, as opportunity for DNA contamination was great from the many people who had handled the mummy. After demonstrating that the DNA they had extracted from the body was not a contaminant, they showed that it matched that of contemporary central and northern Europeans and was distinctly different from Egyptian and South American types. Lingering suspicions that the body was a fraud were therefore squashed.

Dinosaur DNA

Lately, two research groups have claimed that dinosaur DNA may yet be at our fingertips. In July 1993, Jack Horner, of Montana State University, reported the discovery of what appear to be blood cells in the fossilized leg bone of *Tyrannosaurus rex*, perhaps the most famous of all dinosaurs. Horner, a paleontologist, is well respected for his careful and innovative work on dinosaurs. If blood cells are indeed present in the bone, he asked, may it perhaps prove possible to extract DNA from them? So far, none has been found.

In late 1994, researchers at Brigham Young University took the claim a step further, reporting the extraction of DNA from 80-million-year-old dinosaur bone unearthed from a coal mine in eastern Utah. Using PCR, Scott Woodward and his colleagues obtained nine segments of the cytochrome b enzyme of mitochondrial DNA, each measuring more than 130 base pairs. Comparisons of the DNA sequence with modern sequences failed to ally it with any group—mammals, birds, or crocodiles. This was a

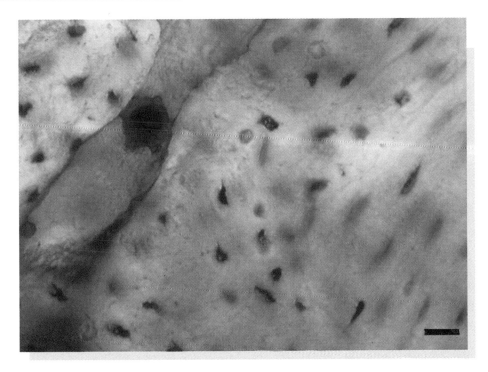

Details of fine structure anatomy are maintained in fossil bone. Here we see a microscopic section of leg bone of *Tyrannosaurus rex*, perhaps the most famous of all dinosaurs, prepared in the Museum of the Rockies, Bozeman, Montana. The small, round microspheres are found within blood vessel channels and have been shown to contain heme, a key component of the blood protein hemoglobin. Because reptilian red blood cells contain nuclei, there was the hope that DNA may also be present in the fossil bone. New research has shown that to be unlikely, however.

surprise, because dinosaur DNA would be expected to share some similarity with that of birds, which are now considered to be descendants of dinosaurs.

A mixture of excitement and skepticism greeted the report, published in the respected journal *Science*. Most observers in the field said they would wait to see if other research groups can replicate the result, by obtaining similar DNA from samples of the same fossilized bone. To some researchers, the claimed distinctiveness of the extracted DNA encouraged the belief that it may indeed be of dinosaur origin; others, such as Blair Hedges of Pennsylvania State University and Mary Schweitzer of the Museum of the Rockies, Bozeman, Montana, suspected it was a contaminant. Six months after Woodward's report appeared, they published their own analysis of the putative dinosaur DNA sequence, and concluded that it was probably human. And exactly a year after Hedges and Schweitzer made their doubts public, in May 1996 *Science* published a research report that all but precludes the possibility that dinosaur DNA will ever be extracted intact from fossil bone. The work implies that even under the most favorable conditions of preservation—that is, in cold climates—DNA may be retrieved from tissue that has been dead for 100,000 years or less. In warm climates, useful quantities of DNA effectively disappear within a few thousand years.

The work, by Svante Pääbo, Hendrik Poinar, and Matthias Höss, of the University of Munich, involves an ingenious way of exploiting a simple chemical reaction of amino acids as an indictor of whether or not a particular tissue sample is likely to contain intact DNA. Amino acids in nature come in two mirror image forms, the D form and the L form, according to their three-dimensional structure. When living systems assemble proteins from a pool of amino acids, they exclusively use the L form. When tissue dies, the protein chains begin to break down, and the L-amino acids begin a slow process of transformation, known as racemization, to the D form, eventually producing an equal mix of the two forms. The dynamics of racemization under various conditions has been studied for some years. In 1993, Jeffrey Bada, of the University of California, San Diego, discovered that the rate of racemization parallels very closely the rate of the principle degradation process of DNA, known as depurination. He therefore joined forces with Pääbo and his colleagues to work out a quantitative measure of racemization as a proxy indicator of the presence of intact DNA.

The researchers chose to measure aspartic acid, because it undergoes racemization faster than all other amino acids. They determined the D and L forms of aspartic acid in tissue ranging in age between 50 and 40,000 years, and they found that when the D/L ratio was 0.08 or higher, no intact DNA remained. They then measured this D/L ratio in various tissues for which claims of ancient-DNA extraction had been made, including the 65-million-year-old *Tyrannosaurus rex* bone, the 17-million-year-old magnolia leaves, and insects in amber of various ages. The D/L ratio exceeded the degradation threshold in all samples, except for the amber-entombed creatures. The researchers concluded that "the prospect of retrieving DNA sequences from dinosaur fossils seems bleak."

In a commentary accompanying the research paper, Scott Woodward conceded that his earlier claims for extraction of dinosaur DNA had been dealt a serious blow. The degradation processes are just too rapacious for long-term preservation of DNA in bone. By contrast, the chemical environment of amber apparently blocks degradation for at least 35 million years (the age of the oldest sample tested). Michael Crichton was therefore right to base his fictional resurrection of dinosaurs on insects in amber.

Even if it did prove possible to extract dinosaur DNA from blood-sucking insects entombed in amber, *Jurassic Park* is still fiction. Dinosaurs are extinct, and likely to remain so. The barriers to "growing" an individual of a living species from raw DNA, as opposed to starting from a full complement of chromosomes safely sequestered inside an egg, would be vast, given the complexity of embryological development and biologists' ignorance of its mechanisms. With extinct species, the fraction of DNA that is rescued from long-dead tissue is minuscule: a handful of fragments 200 base pairs in length, compared with the several billion nucleotides that constitute the full genetic complement in the intact cell. Charles Pellagrino's *Omni* speculation about filling in the missing DNA and inserting it into an egg makes for creative daydreams, but that's all.

When a species becomes extinct, a genetic lineage that stretches back to the beginning of life is lost—forever. Ancient-DNA research can do what a decade ago few thought possible; but it cannot achieve the impossible. Although the resurrection of vanished species remains beyond its powers, ancient-DNA research is part of a revolution in biology that allows wonderful new insights about the living world that still exists—and may help us to find ways of ensuring its continued intricate survival.

Further Readings

CHAPTER 2 MOLECULES VERSUS MORPHOLOGY

Molecules Versus Morphology

Atchley, W. R., and W. M. Fitch. Gene trees and the origin of inbred strains of mice. *Science* 254 (1991): 554–558.

King, M.-C., and A. C. Wilson. Evolution at two levels in humans and chimpanzees. *Science* 188 (1975): 107–116.

Patterson, C., ed. *Molecules and Morphology in Evolution.* Cambridge University Press, 1987.

Methods Compared

Hillis, D. M. Homology in molecular biology. In *Homology: The Hierarchical Basis of Comparative Biology,* B. K. Hall, ed. Academic Press, 1993.

Hillis, D. M., et al. Analysis of DNA sequence data: phylogenetic inference. *Methods in Enzymology* 224 (1993): 456–487.

Patterson, C., et al. Congruence between molecular and morphological phylogenies. *Annual Reviews of Ecology and Systematics,* 24 (1993): 153–188.

de Queriroz, A., et al. Separate versus combined analysis of phylogenetic evidence. Annual Reviews of Ecology and Systematics, 26 (1995): 657–681.

Swofford, D. L. When are phylogeny estimates for molecular and morphological data incongruent? In *Phylogenetic Analysis of DNA Sequences,* M. M. Myamoto and J. Cracraft, eds., 295–333. Oxford University Press, 1994.

Assessing Phylogenetic Methods

Hillis, D. M., et al. Application and accuracy of molecular phylogenies. *Science* 264 (1994): 671–667.

Hillis, D. M., et al. To tree the truth: biological and numerical simulations of phylogeny. In *Molecular Evolution of Physiological Processes,* D. M. Fambrough, ed. The Rockefeller Press, 1994.

Stewart, C.-B. The powers and pitfalls of parsimony. *Nature* 361 (1993): 603–607.

CHAPTER 3 TREES OF LIFE

The Big Tree

Doolittle, R. F., et al. Determining divergence times of the major kingdoms of living organisms with a protein clock. *Science* 271 (1996): 470–477.

Pace, N. Origin of life: facing up to the physical setting. *Cell* 65 (1991): 531–533.

Rivera, M. C., and J. A. Lake. Evidence that eukaryotes and eocyte prokaryotes are immediate relatives. *Science* 257 (1992): 74–76.

Schlegel, M. Molecular phylogeny of eukaryotes. *Trends in Ecology and Evolution* 9 (1994): 330–335.

Woese, C. R. There must be a prokaryote somewhere. *Microbiological Reviews* 58 (1994): 1–9

Woese, C. R., et al. Towards a natural system of organisms. *Proceedings of the National Academy of Sciences,* 87 (1990): 4576–4579.

Early Eukaryotic Phylogenies

Conway Morris, S. The fossil record and the early evolution of the Metazoa. *Nature* 361 (1993): 219–225.

Erwin, D. Metazoan phylogeny and the Cambrian radiation. *Trends in Ecology and Evolution* 4 (1991): 131–134.

Gray, M. W. The endosymbiont hypothesis revisited. *International Reviews of Cytology* 141 (1992): 233–357.

Knoll, A. The early evolution of the eukaryotes: a geological perspective. *Science* 256 (1992): 622–627.

Sidow, A., and W. K. Thomas. A molecular framework for eukaryotic organisms. *Current Biology* 4 (1994): 596–603.

Later Eukaryotic Phylogenies

Ahlberg, P. E., and A. R. Milner. The origin and early diversification of the tetrapods. *Nature* 368 (1994): 507–514.

Bailey, W. J., et al. Rejection of the "flying primate" hypothesis by phylogenetic evidence from the epsilon-globin gene. *Science* 256 (1992): 86–89.

Boore, J. L., et al. Deducing the pattern of arthropod phylogeny from mitochondrial DNA arrangements. *Nature* 376 (1995): 163–167.

Graur, D. Molecular phylogeny and the higher classification of Eutherian mammals. *Trends in Ecology and Evolution* 8 (1993): 141–146.

Meyer, A. Molecular evidence on the origin of tetrapods. *Trends in Ecology and Evolution* 10 (1995): 111–116.

Milinkovitch, M. C. Molecular phylogeny of cetaceans prompts revision of morphological transformations. *Trends in Ecology and Evolution* 10 (1995: 328–334.

Novacek, M. J. Mammalian phylogeny: shaking the family tree. *Nature* 356 (1992): 121–126.

Sibley, C. G., and J. E. Ahlquist. Reconstructing bird phylogeny by comparing DNA's. *Scientific American* (February 1986): 82–92.

Recent Phylogenies

Gargas, A., et al. Multiple origins of lichen symbiosis in fungi suggested by SSU rDNA phylogeny. *Science* 268 (1995): 1492–1495.

Hillis, D. M., et al. Application and accuracy of molecular phylogenies. *Science* 264 (1994): 671–677.

Meyer, A., and A. C. Wilson. Monophyletic origin of Lake Victoria cichlid fish is suggested by mitochondrial DNA sequences. *Nature* 347 (1990): 550–553.

CHAPTER 4 THE PUZZLE OF VARIATION

Extent of Variation

Lewontin, R. C. Electrophoresis in the development of evolutionary genetics: milestone or millstone? *Genetics* 128 (1991): 657–662. (And references therein.)

Origin of Variation

Gillespie, J. H. Molecular evolution and the neutral allele theory. In P. H. Harvey, and L. Partridge, eds., *Oxford Surveys in Evolutionary Biology* 4 (1987): 10–37.

Gould, S. J. Through a lens, darkly. *Natural History* (September 1989): 16–24.

Kimura, M. The neutral theory of molecular evolution: a review of recent evidence. *Japanese Journal of Genetics* 66 (1991): 367–386.

Ohta, T. The nearly neutral theory of molecular evolution. *Annual Reviews of Ecology and Systematics* 23 (1992): 263–286.

CHAPTER 5 THE MOLECULAR EVOLUTIONARY CLOCK

Ayala, F. J. On the virtues and pitfalls of the molecular evolutionary clock. *The Journal of Heredity* 77 (1986): 226–235.

Britten, R. J. Rates of DNA sequence evolution differ between taxonomic groups. *Science* 231 (1986): 1393–1398.

Fitch, W. M., and Ayala, F. J. The superoxide dismutase molecular clock revisited. In *Tempo and Mode in Evolution*, W. M. Fitch, and F. J. Ayala, eds., 235–252. National Academy of Sciences, 1995.

Gillespie, J. H. Natural selection and the molecular clock. *Molecular Biology and Evolution* 3 (1986): 138–155.

Kimura, M. Molecular evolutionary clock and the neutral theory. *Journal of Molecular Evolution* 26 (1987): 24–33.

Martin, A. P., et al. Rates of mitochondrial DNA evolution in sharks are slow compared with mammals. *Nature* 357 (1992): 153–155.

Martin, A. P., and S. R. Palumbi. Body size, metabolic rate, generation time and the molecular clock. *Proceedings of the National Academy of Sciences (USA)* 90 (1993): 4807–4091.

Special issue: Molecular evolutionary clock. *Journal of Molecular Evolution* 26 (1987): 1–171

Zuckerkandl, E. On the molecular evolutionary clock. *Journal of Molecular Evolution* 26 (1987): 34–46.

CHAPTER 6 MOLECULAR ECOLOGY

General

Burke, T. Spots before the eyes: molecular ecology. *Trends in Ecology and Evolution* 9 (1994): 355–357.

Burke, T., et al. Molecular variation and ecological problems. In *Genes in Ecology*, R. J. Berry, et al., eds. 229–254. Blackwell Scientific, 1992.

Evolutionary Ecology

Amos, B., et al. Social structure of pilot whales revealed by analytical DNA profiling. *Science* 260 (1993): 670–672.

Amos, B., et al. Evidence for mate fidelity in the gray seal. *Science* 268 (1995): 1897–1899.

Burke, T. DNA fingerprinting and methods for the study of mating success. *Trends in Ecology and Evolution* 4 (1989): 139–144.

Chapela, I. H., et al. Evolutionary history of the symbiosis between fungus-growing ants and their fungi. *Science* 266 (1994): 1691–1697.

Gibbs, H. L., et al. Realized reproductive success of polygynous red-winged blackbirds revealed by DNA markers. *Science* 250 (1990): 1394–1397.

Mulder, R. A., et al. Helpers liberate female fairy-wrens from constraints on extra-pair mate choice. *Proceedings of the Royal Society (B)*. 255 (1994): 223–229.

Queller, D. C. Microsatellites and kinship. *Trends in Ecology and Evolution* 8 (1993): 285–288.

Richman, A. D., and T. Price. Evolution of ecological differences in the Old World leaf warblers. *Nature* 355 (1992): 817–821.

Behavioral Ecology

Baker, C. S., et al. Mitochondrial DNA variation and world-wide population structure of humpback whales. *Proceedings of the National Academy of Sciences (USA)* 90 (1993): 8239–8243.

Bowen, B. W., and J. C. Avise. Tracking turtles through time. *Natural History* 12/94: 36–42.

Phylogeography

Avise, J. C. Molecular population structure and the biogeographic history of a regional fauna. *Oikas* 63 (1992): 62–76.

Ferris, C., et al. Native oak chloroplasts reveal an ancient divide across Europe. *Molecular Ecology* 2 (1993): 337–344.

Joseph, L. J., et al. Molecular support for vicariance as a source of diversity in rainforest. *Proceedings of the Royal Society (B)* 260 (1995): 177–182.

Conservation Genetics

Avise, J. C. A role for molecular genetics in the recognition and conservation of endangered species. *Trends in Ecology and Evolution* 4 (1989): 279–281.

O'Brien, S. J. Genetic and phylogenetic analyses of endangered species. *Annual Reviews of Genetics* 28 (1994): 467–489.

O'Brien, S. J. A role for molecular genetics in biological conservation. *Proceedings of the National Academy of Sciences (USA)* 91 (1994): 5748–5755.

Daugherty, C. H., et al. Neglected taxonomy and continuing extinctions of tuatara (*Sphenodon*). *Nature* 347 (1990): 177–179.

May, R. M. The cheetah controversy. *Nature* 374 (1995): 309–310.

Menotti-Raymond, M., and S. J. O'Brien. Dating the genetic bottleneck of the African cheetah. *Proceedings of the National Academy of Sciences (USA)* 90 (1993): 3172–3176.

Special issue on conservation genetics. *Molecular Ecology* 3 (1994): 277–435.

Roelke, M. E., et al. The consequences of demographic reduction and genetic depletion in the endangered Florida panther. *Current Biology* 3 (1993): 340–350.

CHAPTER 7 MOLECULAR ANTHROPOLOGY

Human Origins

Bailey, W. Hominoid trichotomy: a molecular overview. *Evolutionary Anthropology* 2 (1993): 100–108. (And references therein.)

Goodman, M. A personal account of the origins of a new paradigm. *Molecular Phylogenetics and Evolution* 5 (1996): 269–285.

Rogers, J. The phylogenetic relationships among *Homo, Pan,* and *Gorilla*: a population genetics perspective. *Journal of Human Evolution* 25 (1993): 201–215.

Ruvolo, M., et al. Gene trees and hominoid phylogeny. *Proceedings of the National Academy of Sciences (USA)* 91 (1994): 8900–8904.

Special issue on molecular anthropology. *Molecular Phylogenetics and Evolution* 5 (1996): 1–285.

White, T. D., et al. *Australopithecus ramidus*, a new species of early hominid from Aramis, Ethiopia. *Nature* 371 (1994): 306–329.

Wood, B. The oldest hominid yet. *Nature* 371 (1994): 280–281.

Origin of Modern Humans

Bowcock, A. M., et al. High resolution of evolutionary trees with polymorphic microsatellites. *Nature* 368 (1994): 455–457.

Cavalli-Sforza, L. L. Genes, people and languages. *Scientific American* (November 1991): 104–110.

Cavalli-Sforza, L. L., et al. Demic expansion and human evolution. *Science* 259 (1993): 639–646.

Dorit, R. L., et al. Absence of polymorphism at the ZFY locus on the human Y chromosome. *Science* 268 (1995): 1183–1185.

Lahr, M. M., and R. Foley. Multiple dispersals and modern human origins. *Evolutionary Anthropology* 3 (1994): 48–60. (And references therein.)

Ruvolo, M. A new approach to studying modern human origins. *Molecular Phylogenetics and Evolution* 5 (1996): 202–219.

Stoneking, M. DNA and recent human evolution. *Evolutionary Anthropology* 2 (1993): 60–73. (And references therein.)

Stringer, C. B. The emergence of modern humans. *Scientific American* (December 1990): 98–104.

Thorne, A. G., and M. H. Wolpoff. The multiregional evolution of humans. *Scientific American* (April 1992): 76–83.

Tishkoff, S. A., et al. Global patterns of linkage disequilibrium at the CD4 locus and modern human origins. *Science* 271 (1996): 1380–1387.

Wilson, A. C., and R. L. Cann. The recent African genesis of humans. *Scientific American* (April 1992): 68–73.

Peopling of the Americas

Hoffecker, J. F., et al. The colonization of Beringia and the peopling of the New World. *Science* 259 (1993): 46–53.

Meltzer, D. J. Why don't we know when the first people came to North America? *American Antiquity* 54 (1989): 471–490.

Meltzer, D. J. Pleistocene people of the Americas. *Evolutionary Anthropology* 1 (1993): 157–169. (And references therein.)

Merriweather, D. A., and R. E. Ferrel. The four founding lineage hypothesis for the New World. *Molecular Phylogenetics and Evolution* 5 (1996): 241–246.

Szathmary, E. J. E. Genetics of aboriginal North Americans. *Evolutionary Anthropology* 1 (1993): 202–220. (And references therein.)

Torroni, A., et al. Mitochondrial DNA "clock" for the Amerinds and its implications for timing their entry into North America. *Proceedings of the National Academy of Sciences (USA)* 91 (1994): 1158–1162.

Ward, R. H., et al. Genetic and linguistic differentiation in the Americas. *Proceedings of the National Academy of Sciences (USA)* 90 (1993): 10,663–10,667.

Weiss, K. M. American origins. *Proceedings of the National Academy of Sciences (USA)* 91 (1994): 833–835.

CHAPTER 8   ANCIENT DNA

Brown, T. A., and K. A. Brown. Ancient DNA and the archeologist. *Antiquity* 66 (1992): 10–23.

DeSalle, R., et al. DNA sequences from a fossil termite in Oligo-Miocene Amber and their phylogenetic implications. *Science* 257 (1992): 1933–1936.

Gibbons, A. Possible dino DNA find is greeted with skepticism. *Science* 266 (1994): 1159.

Hagelberg, E. Ancient DNA studies. *Evolutionary Anthropology* 2 (1993): 199–206.

Handt, O. Molecular genetic analyses of the Tyrolean Ice Man. *Science* (1994): 1775–1778.

Hardy, C., et al. Origin of the European rabbit (*Oryctolagus cuniculus*) in a Mediterranean island: zoogeography and ancient DNA examination. *Journal of Evolutionary Biology* 7 (1994): 217–226.

Hauswirth, H. Dead men's molecules. *Yearbook of Science and Technology,* 122–137. Encyclopedia Britannica, 1994.

Lewin, R. Fact, fiction and fossil DNA. *New Scientist* (29 January 1994): 38–41.

Lindahl, T. Instability and decay of the primary structure of DNA. *Nature* 362 (1993): 709–715.

Pääbo, S. Ancient DNA. *Scientific American* (November 1993): 60–66.

Poinar, G. O., Jr. Still life in amber. *The Sciences* (March/April 1993): 34–38.

Poinar, H. N., et al. Amino acid racemization and the preservation of ancient DNA. *Science* 272 (1996): 864–866.

Ross, P. E. Eloquent Remains. *Scientific American* (May 1992): 115–125.

Sjovold, T. The Stone Age iceman from the Alps. *Evolutionary Anthropology* 1 (1992): 117–124.

Thomas, R. H., et al. DNA phylogeny of the extinct marsupial wolf. *Nature* 340 (1989): 465–467.

Woodward, S. R., et al. DNA sequence from Cretaceous period bone fragments. *Science* 266 (1994): 1229–1232.

Sources of Illustrations

Chapter 1 Facing page 1: © Rosamond Purcell. All rights reserved Page 2: D. N. Dalton/Natural History Photo Agency Page 3: James Holmes/CellMark Diagnostics/Photo Researchers Page 5: Reproduced by permission of The Linnaean Society of London Page 6: Photograph by Julia Margaret Cameron, 1868. National Portrait Gallery, London Page 10: © Rosamond Purcell. All rights reserved Page 16: Benali Landmann/Gamma Liaison **Chapter 2** Page 18: © Rosamond Purcell. All rights reserved Page 21: Oil painting by H. W. Pickersgill, 1845. National Portrait Gallery, London Page 22: Adapted from a figure in Charles G. Sibley and Jon E. Ahlquist, Reconstructing bird phylogeny by comparing DNA's, *Scientific American* 254(2): 84. © 1986 by Scientific American, Inc. All rights reserved Page 27: Wolfgang Hennig Page 32: John Reader/Photo Researchers Page 49: From Roger Lewin, *Human Evolution*, 3rd ed., Blackwell Scientific Publications, 1993 Page 50: M. E. Bisher and A. C. Steven Page 51: From J. J. Bull et al., Experimental molecular evolution of bacteriophage T7, *Evolution* 47 (1993): 993–1007 **Chapter 3** Page 52: © Rosamond Purcell. All rights reserved Page 59: Based on a figure in C. R. Woese et al., Towards a natural system of organisms, *Proceedings of the National Academy of Sciences* 87 (1990): 4576–4579 Page 60: Photograph by Thomas Brock, University of Wisconsin at Madison. Courtesy of University of Wisconsin Office of News and Public Affairs Page 61: Reg Morrison/AUSCAPE Page 62: Adapted from a figure in M. Schlegel, Molecular phylogeny of eukaryotes, *Trends in Ecology and Evolution* 9 (1994): 331 Page 67: From A. Knoll, The early evolution of the eukaryotes: a geological perspective, *Science* 256 (1992): 623 Page 70: From K. G. Field et al., Molecular phylogeny of the animal kingdom, *Science* 239 (1988): 748–753 Page 73: Peter Scoones/Planet Earth Pictures Page 75: From D. Graur, Molecular phylogeny and the higher classification of Eutherian mammals, *Trends in Ecology and Evolution* 8 (1993): 141–146 Page 76: Stephen Dalton/Natural History Photo Agency Page 78: Jean-Paul Ferrero/AUSCAPE Page 79: From C. G. Sibley and J. E. Ahlquist, Reconstructing bird phylogeny by comparing DNA's, *Scientific American* 254(2): 82–92. © 1986 by Scientific American, Inc. All rights reserved Page 80: Hans and Judy Beste/AUSCAPE Page 82: Photograph by Andreas Spreinat. Courtesy of Axel Meyer, University of California at Berkeley **Chapter 4** Page 86: Peter Ginter/Bilderberg Page 91: Rockefeller University Archive Center Page 92: Indiana University Archives Page 97: William B. Provine, Cornell University Page 102: From Motoo Kimura, The neutral theory of molecular evolution, *Scientific American* 241(5). © 1979 by Scientific American, Inc. All rights reserved Page 103: Eviatar Nevo, Institute of Evolution, University of Haifa **Chapter 5** Page 106: Frans Lanting/Minden Pictures Page 110 (*left*): Emile Zuckerkandl Page 110 (*right*): Janet Fries/Black Star Page 111: Chip Clark Page 112: Jane Scherr, Wilson Laboratory, University of California at Berkeley Page 113: Vincent Sarich **Chapter 6** Page 120: Wolfgang Volz/Bilderberg Page 123: From J. D. Watson, M. Gilman, J. Witkowski, and M. Zoller, *Recombinant DNA*, 3rd ed. © 1992 by Scientific American Books. All rights reserved Page 126: William Taufic Photography, Inc. Page 128: Michio Hoshino/Minden Pictures Page 130 (*left*): Robert Erwin/Natural History Photographic Agency Page 130 (*right*): Art Wolfe Page 131: A.N.T./Natural History Photographic Agency Page 133: Margaret Welby/Planet Earth Pictures Page 134: Doug Perrine/Planet Earth Pictures Page 136: Madhusudan Katti/Trevor Price Page 137: Mark W. Moffett/Minden Pictures Page 139: Haroldo Palo Jr./Natural History Photographic Agency Page 142: Pete Atkinson/Planet Earth Pictures Page 147: Georgette Douwma/Planet Earth Pictures Page 151: Frans Lanting/Minden Pictures Page 153: Art Wolfe Page 155: Art Wolfe Page 160: Fotocentre/Natural History Photographic Agency **Chapter 7** Page 162: © Rosamond Purcell. All rights reserved Page 166: Morris Goodman Page 169: John Reader/SPL/Photo Researchers Page 170: © 1995 David L. Brill/Atlanta Page 176: Douglas Wallace, Emory University Page 178: Adapted from a figure in Allan C. Wilson and Rebecca L. Cann, The recent African genesis of humans, *Scientific American* 266(4): 69. © 1992 by Scientific American, Inc. All rights reserved Page 179: Adapted from a figure in Allan C. Wilson and Rebecca L. Cann, The recent African genesis of humans, *Scientific American* 266(4): 72. © 1992 by Scientific American, Inc. All rights reserved Page 181: Ira Block Photography, Ltd. Pages 185 and 186: Alison S. Brooks, Department of Anthropology, George Washington University Page 192: Engraving by Theodore De Bry, 16th century. Rare Book and Special Collections Division, Library of Congress Page 194: From Joseph H. Greenberg and Merritt Ruhlen, Linguistic origins of Native Americans, *Scientific American* 267(5): 95. © 1992 by Scientific American, Inc. All rights reserved Page 197: Rick Wicker, Denver Museum of Natural History Photo Archives. All rights reserved **Chapter 8** Page 200: © Rosamond Purcell. All rights reserved Page 202: Benali-Landmann/Gamma Liaison Page 206: Thomas Stephan/Black Star Page 209: Philip Saltonstal Page 212: Photograph by Jacky Beckett. © American Museum of Natural History Page 217: Benali-Landmann/Gamma Liaison Page 221: Alex Webb/Magnum Page 223: Photograph by J. Nauta. The Museum of New Zealand Te Papa Tongarewa, Wellington, New Zealand Pages 227 and 228: Wolfgang Neeb, University of Innsbruck Stern/Black Star Page 231: Photograph by Mary Higby Schweitzer, Montana State University. Source of bone tissue: Museum of the Rockies

Index

Selected Books in the Scientific American Library Series

POWERS OF TEN
by Philip and Phylis Morrison and the Office
of Charles and Ray Eames

ISLANDS
by H. William Menard

DRUGS AND THE BRAIN
by Solomon H. Snyder

EYE, BRAIN, AND VISION
by David H. Hubel

ANIMAL NAVIGATION
by Talbot H. Waterman

SLEEP
by J. Allan Hobson

FROM QUARKS TO THE COSMOS
by Leon M. Lederman and David N. Schramm

SEXUAL SELECTION
by James L. Gould and Carol Grant Gould

THE NEW ARCHAEOLOGY AND THE ANCIENT MAYA
by Jeremy A. Sabloff

A JOURNEY INTO GRAVITY AND SPACETIME
by John Archibald Wheeler

BEYOND THE THIRD DIMENSION
by Thomas F. Banchoff

DISCOVERING ENZYMES
by David Dressler and Huntington Potter

THE SCIENCE OF WORDS
by George A. Miller

ATOMS, ELECTRONS, AND CHANGE
by P. W. Atkins

VIRUSES
by Arnold J. Levine

DIVERSITY AND THE TROPICAL RAINFOREST
by John Terborgh

STARS
by James B. Kaler

EXPLORING BIOMECHANICS
by R. McNeill Alexander

CHEMICAL COMMUNICATION
by William C. Agosta

GENES AND THE BIOLOGY OF CANCER
by Harold Varmus and Robert A. Weinberg

SUPERCOMPUTING AND
THE TRANSFORMATION OF SCIENCE
by William J. Kaufmann III and Larry L. Smart

MOLECULES AND MENTAL ILLNESS
by Samuel H. Barondes

EXPLORING PLANETARY WORLDS
by David Morrison

EARTHQUAKES AND GEOLOGICAL DISCOVERY
by Bruce A. Bolt

THE ORIGIN OF MODERN HUMANS
by Roger Lewin

THE EVOLVING COAST
by Richard A. Davis, Jr.

THE LIFE PROCESSES OF PLANTS
by Arthur W. Galston

IMAGES OF MIND
by Michael J. Posner and Marcus E. Raichle

THE ANIMAL MIND
by James L. Gould and Carol Grant Gould

MATHEMATICS: THE SCIENCE OF PATTERNS
by Keith Devlin

A SHORT HISTORY OF THE UNIVERSE
by Joseph Silk

THE EMERGENCE OF AGRICULTURE
by Bruce D. Smith

ATMOSPHERE, CLIMATE, AND CHANGE
by Thomas E. Graedel and Paul J. Crutzen

AGING: A NATURAL HISTORY
by Robert E. Ricklefs and Caleb Finch

INVESTIGATING DISEASE PATTERNS:
THE SCIENCE OF EPIDEMIOLOGY
by Paul D. Stolley and Tamar Lasky

GRAVITY'S FATAL ATTRACTION:
BLACK HOLES IN THE UNIVERSE
by Mitchell Begelman and Martin Rees

CONSERVATION AND BIODIVERSITY
by Andrew P. Dobson

PLANTS, PEOPLE, AND CULTURE:
THE SCIENCE OF ETHNOBOTANY
by Michael J. Balick and Paul Alan Cox

LIFE AT SMALL SCALE: THE BEHAVIOR OF MICROBES
by David B. Dusenbery